Computer Algorithms: Introduction to Design and Analysis

Computer Algorithms: Introduction to Design and Analysis

Sara Baase
San Diego State University

714 – 286 5200
5231

▲ *ADDISON-WESLEY*
PUBLISHING
COMPANY
Reading, Massachusetts
Menlo Park, California
London
Amsterdam
Don Mills, Ontario
Sydney

This book is in the
Addison-Wesley Series in Computer Science.

Consulting editor
Michael A. Harrison

ISBN 0-201-00327-9
ABCDEFGHIJK-HA-798

To Jack

For never locking out interrupts

Preface

This book is intended for a one-semester, upper-division course. It has sufficient material to allow several choices of topics and algorithms for the latter part of the semester. Familiarity with data structures such as linked lists, stacks, and trees is assumed. Recursion is used in a few algorithms and a very small amount of calculus is used.

The purpose of the book is two-fold. It is intended both to teach algorithms for solving real problems that arise frequently in computer applications and to teach basic principles and techniques for analyzing algorithms. At least as important as teaching the subject matter is another of the book's aims—to develop in the reader the habit of always responding to a new algorithm with the questions: How good is it? and Is there a better way? Therefore, instead of presenting a series of complete, "pulled-out-of-a-hat" algorithms with analysis, the text discusses a problem first, considers one or more approaches to solving it (as a reader who sees the problem for the first time might), and then begins to develop an algorithm, analyzes it, and modifies or rejects it until a satisfactory result is produced. Questions such as: How can this be done more efficiently? What data structure would be useful here? How must this variable (or data structure) be initialized? and so on appear frequently throughout the text. Answers generally follow the question, but I suggest that the reader pause before reading the ensuing text and think up his or her own answers. Learning is not a passive process.

The main points covered in the analyses are the amount of work done (both worst case and average), space usage, and lower bounds on the complexity of a problem. I hope that the reader will also learn to be aware of how an algorithm actually behaves on various inputs—that is, Which tests are successful? In what sequence are the sections of the algorithm executed? What is the pattern of growth and shrinkage of stacks? How do the data structures look at intermediate steps? How does presenting an input in different ways (e.g., inputting the vertices or edges of a graph in different orders) affect the behavior? Such questions are raised in some of the exercises, but are not emphasized in the text because they require carefully going through the details of many examples. The reader should keep them in mind.

Most of the algorithms presented are of practical use; I have chosen not to emphasize those with good asymptotic behavior that are poor for inputs of useful sizes (though some of these, which are of theoretical interest, are mentioned). Specific algorithms were chosen for a variety of different reasons including the importance of the problem, showing a variety of approaches to one problem, illustrating analysis techniques, illustrating techniques (e.g., depth-first search) that give rise to algorithms for numerous problems, and illustrating the development and improvement of techniques and algorithms (e.g., UNION-FIND programs). (Section 2.9 on external sorting was included at the request of many of our working students who said it would be very useful for their work.) The frequent occurrence of problems of very high complexity motivated the inclusion of Chapter 7. Here the reader is introduced to $\mathcal{NP}$-completeness and approximation algorithms for combinatorial problems.

A one-semester course might contain the following material:

Chapter 1, Sections 1, 3, and 4 (with Section 2 on the algorithm language assigned reading);

Chapter 2, Sections 1-5, 8, and one of the remaining sections;

Chapter 3 (perhaps omitting the last section);

Chapter 4;

Chapter 5, Sections 1-3;

Chapter 7, Sections 1, 2, and 4;

Some of the omitted sections or some of Chapter 6, if time allows.

The exercises are an important part of this book, as they should be with any text. The reader should read them all and work on as many as time and interest indicate. Some of the exercises are somewhat open-ended; for example, one might ask for a good lower bound for the complexity of a problem rather than asking the reader to show that a given function is a lower bound. The intent is to make the form of the question more realistic; a solution must be discovered as well as verified.

Algorithms are described in a high-level language using IF-THEN-ELSE, WHILE, FOR, and other common programming language statements as well as instructions in English. The intent is to make the algorithms clear without worrying about language details. Although actual programs are not given and analyzed in the text, data structures and implementation details are considered and discussed at length where these are particularly complex or tricky or are critical to the analysis, that is, where different implementations would result in significantly different complexities for an algorithm.

This book concentrates on and analyzes algorithms rather than programs for several reasons. One is that a good algorithm, or method of solution, is a necessary ingredient for a good program; it is the fundamental thing. A program will be no better than the algorithm and it will contain many details not relevant to the method and the analysis. Of course, the student must learn how to produce programs that are *as* good as the algorithm. Hence another reason for not providing programs in the text is that I expect the students to write them. Each chapter (except the last) includes a list of programming assignments to give the student practice in implementing algorithms. Some of the assignments require a small amount of implementation-dependent analysis. I suggest that the instructor select a few of the algorithms covered in class and do careful presentations and analyses of programs for them in the programming language used by the students.

The bibliography contains a small number of references to relevant papers and books. I have not attempted to include all the original sources of the material in the book because it does not seem necessary or very useful for a book at this level. If there is one source for a major algorithm or section, however, it is included. Generally I have tried to include enough so that the motivated student will have a starting point for further reading.

I want to express my appreciation to my students of the past few years who put up with several early versions of the book with their many errors, typos, and obscurities; to Paul Young and Mike Machtey for using early versions of the manuscript in their classes and providing encouragement from the early months of the project; to the books of Donald Knuth and the lectures of Richard Karp which rekindled my interest in computer science; and, most of all, to Jack Revelle who taught me much of the material in this book and who also taught me how to learn.

San Diego   S. B.
June 1978

Contents

"HARD" ($\mathcal{NP}$-COMPLETE) PROBLEMS AND   7
APPROXIMATION ALGORITHMS

Data Structures and Mathematical Background

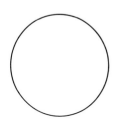

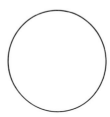

0.1 *Miscellaneous Notation and Formulas*

$\lvert S \rvert$	The cardinality of (i.e., the number of elements in) S, where S is a finite set
$\lvert x \rvert$	The length of (i.e., the number of characters in) x, where x is a character string
$O(f(n))$	Order at most $f(n)$; see Section 1.3
$\Theta(f(n))$	Order $f(n)$; see Section 1.3
$\lfloor x \rfloor$	"Floor of x," the greatest integer $\leqslant x$
$\lceil x \rceil$	"Ceiling of x," the least integer $\geqslant x$
$\log n$	$\log_2 n$
$\ln x$	The natural logarithm of x, i.e., log to the base e
Λ	The null pointer
$\approx$	Approximately equal
∞	Infinity or, when it appears as the value of a variable or a field in a data structure, a number higher than any that would occur in that context
$\binom{n}{k}$	Binomial coefficient: $\binom{n}{k} = \dfrac{n!}{k!(n-k)!}$

0.2 *Some Mathematical Background*

A variety of mathematical concepts such as logarithms, probability, and permutations are used in this book. Most of the time only an understanding of the definition and some elementary properties of these concepts are needed. The purpose of this section is to provide a brief review and reference for this mathematical material, some of which the reader should already have been exposed to in other classes.

LOGARITHMS

The most extensively used mathematical tool in this book is the logarithm function to the base 2.

Definition For $x > 0$, $\log x$ is that number y such that $2^y = x$; i.e., $\log x$ is the power to which 2 must be raised to get x.

The following properties follow easily from the definition (and all but Property 4 are valid for logarithm functions for bases other than 2). In all cases, x's are arbitrary positive numbers and a is any real number.

1. log is a strictly increasing function; i.e., if $x_1 > x_2$, then $\log x_1 > \log x_2$.
2. log is a one-to-one function; i.e., if $\log x_1 = \log x_2$, then $x_1 = x_2$.
3. $\log 1 = 0$.
4. $\log 2^a = a$. (For another base b, use $\log x = \log_b x \log b$.)
5. $\log (x_1 x_2) = \log x_1 + \log x_2$.
6. $\log (x^a) = a \log x$
7. $x_1^{\log x_2} = x_2^{\log x_1}$. (To prove this, show that the logs of both sides of the equation are equal.)

Throughout the text we almost always use logs of integers only, not arbitrary positive numbers, and we often need an integer value close to the log rather than its exact value. Let n be a positive integer. If n is a power of 2, say $n = 2^k$, for some integer k, then $\log n = k$. If n is not a power of 2, then there is an integer k such that $2^k < n < 2^{k+1}$, and $\lfloor \log n \rfloor$, the greatest integer $\leqslant \log n$, is k, and $\lceil \log n \rceil$, the smallest integer $\geqslant \log n$, is $k + 1$. $\lfloor \log n \rfloor$ and $\lceil \log n \rceil$ are used often. The reader should verify that $n \leqslant 2^{\lceil \log n \rceil} < 2n$ and $n/2 < 2^{\lfloor \log n \rfloor} \leqslant n$.

PROBABILITY

Suppose that in a given situation any one of, say, k events $s_1, s_2, \ldots, s_k$ may occur. With each event s_i we associate a number $p(s_i)$, called the probability of s_i, such that

1. $0 \leqslant p(s_i) \leqslant 1$ for $1 \leqslant i \leqslant k$, and
2. $p(s_1) + p(s_2) + \cdots + p(s_k) = 1$.

It is natural to interpret $p(s_i)$ as the ratio of the number of times s_i is expected to occur to the total number of times the situation occurs. If $p(s_i) = 0$, then s_i is impossible; if $p(s_i) = 1$, then s_i always occurs. (Note, however, that the definition does not require that the probabilities assigned correspond to anything in the real world.)

Probably the most frequently used examples to illustrate the meaning of probability are flipping coins and throwing dice. If the situation to be observed (often called the "experiment") is the flip of a coin, then the coin may land with "heads" facing up or with "tails" facing up. We let $s_1 = $ 'heads' and $s_2 = $ 'tails' and assign $p(s_1) = \frac{1}{2}$ and $p(s_2) = \frac{1}{2}$. (If the reader objects because the coin could land on its edge, we may let $s_3 = $ "edge" and define $p(s_3) = 0$.) If a six-sided die is thrown, there are six possible events: for $1 \leqslant i \leqslant 6$, $s_i = $ "the die lands with side number i facing up," and $p(s_i) = \frac{1}{6}$. In general, if there are k events, each considered equally likely to occur, we let $p(s_i) = 1/k$ for each i.

It is often useful to speak of the probability of any one of several specified events occurring or the probability that the event that occurs has a particular property. Let S be a subset of $\{s_1, \ldots, s_k\}$. Then $p(S) = \sum_{s_i \in S} p(s_i)$. For example, the probabilty that when a die is thrown the number appearing is divisible by 3 is $p(\{s_3, s_6\}) = p(s_3) + p(s_6) = \frac{1}{3}$.

PERMUTATIONS

A permutation of n objects is, informally speaking, a rearrangement of the objects. Let $S = \{s_1, s_2, \ldots, s_n\}$. Note that the elements of S are ordered by their indexes; i.e., s_1 is the first element, s_2 the second, and so on. A permutation of S is a one-to-one function π from the set $\{1, 2, \ldots, n\}$ onto itself. We think of π as rearranging S by moving the ith element, s_i, to the $\pi(i)$th position. We may describe π simply by listing its values, that is, $(\pi(1), \pi(2), \ldots, \pi(n))$. For example, for $n = 5$, $\pi = (4, 3, 1, 5, 2)$ rearranges the elements of S as follows: s_3, s_5, s_2, s_1, s_4.

The number of permutations of a set with n elements given in a particular order, $s_1, s_2, \ldots, s_n$, is $n!$. To see this, observe that the first element can be moved to any one of the n positions; then that position is filled and the second element can be moved to any of the $n - 1$ remaining positions; the third element can be moved to any of the remaining $n - 2$ positions, and so on. So the total number of possible rearrangements is $n \times (n - 1) \times (n - 2) \times \cdots \times 2 \times 1 = n!$.

SUMMATION FORMULAS

There are several summations that occur frequently when analyzing algorithms. Formulas for a few of these are listed here with brief hints that may help the reader to remember them.

1.
$$\sum_{i=1}^{n} i = \frac{n(n + 1)}{2}$$

How to remember it. Write out the integers from 1 to n. Pair up the first and last, i.e., 1 and n; pair up the second and next to last, 2 and $n - 1$, etc. Note that each pair adds up to $n + 1$. If n is even, there are $n/2$ pairs; hence the sum is $(n/2)(n + 1)$. If n is odd, there are $(n - 1)/2$ pairs and the extra number $(n + 1)/2$ left in the middle; thus the sum is

$$\frac{n-1}{2}(n+1) + \frac{n+1}{2} = \frac{n}{2}(n+1).$$

2.
$$\sum_{i=0}^{k} 2^i = 2^{k+1} - 1$$

How to remember it. Think of each term 2^i as a one-bit in a binary number:

$$\sum_{i=0}^{k} 2^i = \underbrace{1\,1\ldots 1}_{k+1\ \text{1's}}.$$

If 1 is added to this number the result is

$$\underbrace{1\,0\,0\ldots 0}_{k+1\ \text{0's}} = 2^{k+1}.$$

(This result can also be obtained by using the formula for the sum of a geometric progression given in (3) below.)

3.
$$\sum_{i=0}^{k} \frac{1}{2^i} = 2 - \frac{1}{2^k}$$

This is a geometric progression. The general form and the formula for the sum is

$$\sum_{i=0}^{k} a^i = \frac{a^{k+1} - 1}{a - 1}.$$

To verify this, divide out the right-hand side.

0.3   *Data Structures*

We assume that the reader is already familiar with data structures such as linked lists, stacks, queues, and binary trees. The point of this section is to briefly describe the structures to be used in this book and their terminology.

The nodes, or individual elements, of a data structure may be divided into several *fields*. Each field is given a name. For example the node format for a linked list may be

NAME	AGE	LINK

where the NAME field contains a character string, the AGE field contains a positive integer, and the LINK field contains a pointer. A *pointer* (or *link*) is a variable, array element, or field entry that "points to" a node; in other words, its value is an address (or array index) indicating the location of a node. Pointers are drawn as arrows in diagrams of data structures. (The value of a pointer always indicates the location of the beginning of a node, even though in the diagrams the arrow may point at some other part of the node – e.g., the end – to keep the picture clear.) The *null pointer* is a pointer value that indicates that there is no node pointed to; it will be denoted by Λ.

When using linked data structures we refer to a particular field of a particular node by specifying the field name and a pointer to the node. The pointer is enclosed in parentheses. For example, if *P* is a pointer variable, LINK(*P*) refers to the LINK field in the node pointed to by *P*. (If the pointer value is Λ, this notation is undefined.) More examples are given below.

A *linked list* is a finite sequence of nodes such that each has a pointer field containing a pointer to the next node. A pointer to the list, i.e., to the first node, must be given. The rest of the description of the structure depends on the kind of linked list it is. In a *simply linked list* the pointer field in the last node contains Λ, and if the list is empty the pointer to the list has value Λ. In the following example of a simply linked list, the node format is

and FIRST is a pointer variable.

Using the notation described above for field entries, NAME(FIRST) = 'ERIC', AGE(LINK(FIRST)) = 12, and LINK(LINK(LINK(FIRST))) = Λ. If *P* is another pointer variable whose current value is the location of the third node, then NAME(*P*) = 'ROBIN' and LINK(*P*) = Λ.

A *listhead* is a special node that may be included at the beginning of a linked list; it contains a pointer to the first data node and possibly some other information about the list (for example, the number of data nodes it contains). If a listhead is used, it is never deleted from the list; thus its use simplifies some of the details of inserting and deleting nodes by eliminating the special case of an empty list.

A *circular linked list* is a linked list in which the last node contains a pointer to the first node (or listhead, if there is one) instead of Λ.

A *doubly linked list* is a linked list in which each node contains a pointer to the previous node, as well as to the next one. A doubly linked list may be circular and may have a listhead. The use of both of these features eliminates the special cases that occur when inserting or deleting nodes at either end of the list and thus simplifies algorithms. In the following example of a doubly linked circular list with listhead, the node format is

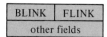

where BLINK is the back link field and FLINK is the forward link field.

LIST

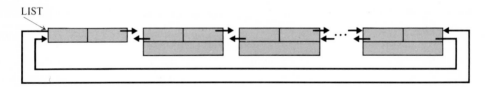

A *stack* is a list in which insertions and deletions are always made at one end, called the *top*. The top item in a stack is the one most recently inserted. A stack may be implemented or stored in an array or as a linked list. To *push* an item on a stack means to insert the item in the stack. To *pop* the stack means to delete the top entry. Stacks are sometimes called LIFO lists ("last in, first out").

A *queue* is a list in which all insertions are done at one end, called the *rear* or *back*, and all deletions are done at the other end, called the *front*. A queue may be stored in an array or as a linked list. Queues are sometimes called FIFO lists ("first in, first out").

A *binary tree* is a finite set of elements, called nodes, which is empty or else satisfies the following:

1.   There is a distinguished node called the *root*, and

2.   The remaining nodes are divided into two disjoint subsets, L and R, each of which is a binary tree. L is the *left subtree* of the root and R is the *right subtree* of the root.

Binary trees are represented on paper by diagrams such as the one in Fig. 0.1. If a node w is the root of the left (right) subtree of a node v, w is the *left (right) child* of v

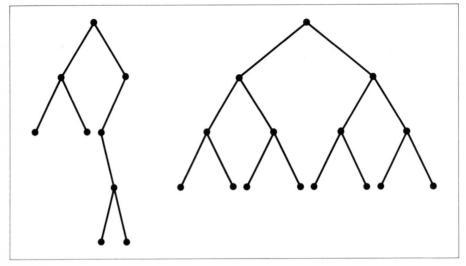

Figure 0.1
Binary trees.

and v the *parent* of w; there is a branch connecting v and w in the diagram. The *degree* of a node is the number of nonempty subtrees it has. A node with degree zero is a *leaf*. Noted with positive degree are *internal nodes*. The *level* of the root is 0 and the level of any other node is one plus the level of its parent.* The *depth* (sometimes called the *height*) of a binary tree is the maximum of the levels of its leaves. A *complete binary tree* is a binary tree in which all internal nodes have degree 2 and all leaves are at the same level. The second binary tree in Fig. 0.1 is complete.

A binary tree is usually represented in the computer as a linked structure: The location of the root is specified by a pointer variable and each node has two pointer fields, LCHILD and RCHILD, which contain pointers to its left child and right child, respectively. Pointers to the parent of each node may be used also or instead if they are needed.

The following facts are used often in the text. The proofs are easy and are omitted.

Lemma 0.1 There are at most 2^ℓ nodes at level ℓ of a binary tree.

Lemma 0.2 A binary tree with depth d has at most $2^{d+1} - 1$ nodes.

Lemma 0.3 A binary tree with n nodes has depth at least $\lfloor \log n \rfloor$.

A *tree* (sometimes called a general tree to distinguish it from a binary tree) is a finite nonempty set of nodes satisfying the following:

1.   There is a distinguished node called the root, and

2.   The remaining nodes are partitioned into $m \geq 0$ disjoint subsets, $T_1, T_2, \ldots, T_m$, each of which is a tree. $T_1, \ldots, T_m$ are called the subtrees of the root.

The terms parent, child, leaf, degree, level, and depth are defined for trees as they are for binary trees with minor modifications.

A *forest* is a finite (possibly empty) set of trees.

NOTES AND REFERENCES

Horowitz and Sahni (1976) is an excellent text to use as a reference for data structures. (References are listed at the end of the book.)

---

*   Beware: Some authors define level so that the level of the root is 1.

Analyzing Algorithms: Principles and Examples

1.1   *Introduction*

To say that a problem is solvable algorithmically means, informally, that there is a computer program that will produce the correct answer for any input if we let it run long enough and allow it as much storage space as it needs. In the 1930s, before the advent of computers, mathematicians worked very actively to formalize and study the notion of an algorithm, which was then interpreted informally to mean a clearly specified set of simple instructions to be followed to solve a problem or compute a function. Various formal models of computation were devised and investigated. Much of the emphasis in the early work in this field, called computability theory, was on describing or characterizing those problems that could be solved algorithmically and on exhibiting some problems that could not be. One of the important negative results was the proof of the unsolvability of the "halting problem." The halting problem is to determine whether an arbitrary given algorithm (or computer program) will get into an infinite loop while working on a given input. There cannot exist a computer program that solves this problem.

Although computability theory has obvious and fundamental implications for computer science, the knowledge that a problem can theoretically be solved on a computer is not sufficient to tell us whether it is practical to do so. For example, a perfect chess playing program could theoretically be written, and it would not be a very difficult task. There is only a finite number of ways of arranging the chess pieces on the board and under certain rules a game must terminate after a finite number of moves. The program could consider each of the computer's possible moves, each of its opponent's possible responses, each of its possible responses to those moves, and so on until each sequence of possible moves reaches an end. Then since it knows the ultimate results of each move, the computer can choose the best one. Such a program has not been run because the number of sequences of reasonable moves (at least 10^{19} by some estimates) is so large that a program which made a thorough examination of them would take several thousand years to execute.

There are numerous problems with practical applications that can be solved — that is, for which programs can be written — but for which the time and storage requirements are much too great for these programs to be of practical use. Clearly the time and space requirements of a program are of practical importance. They have become, therefore, the subject of theoretical study in the area of computer science called computational complexity. One branch of this study, which is not covered in this book, is concerned with setting up a formal and somewhat abstract theory of the complexity of computable functions. (Solving a problem is equivalent to computing a function from the set of inputs to the set of outputs.) Axioms for measures of complexity have been formulated and are basic and general enough so that either the number of instructions executed or the number of storage bits used by a program can be taken as a complexity measure. Using these axioms, one can prove the existence of arbitrarily complex problems and of problems for which there is no best program.

The branch of computational complexity studied in this book is concerned with analyzing specific problems and specific algorithms. The term "complexity" refers to the time or space requirements of a problem or algorithm. We will consider a number of problems for which computer programs are frequently used, and will examine and analyze algorithms to solve them. This book is intended to help the reader build up a repertoire of efficient algorithms to solve common problems and to provide the tools and principles of computational complexity for analyzing and improving algorithms.

The questions below motivate much of this book. Answers are given, suggested, and/or illustrated throughout the book but are not specifically identified as such. The reader is encouraged to return to these questions occasionally, perhaps at the end of each chapter, and think out or write down answers.

How can given algorithms and programs be improved?

How can the efficiency of algorithms be analyzed mathematically?

What criteria may be used to choose between different algorithms for the same application?

In what sense can algorithms be shown to be the best possible?

How useful are the formal definitions and theoretical results in practice?

1.2   *The Algorithm Language*

The algorithm language used in this book was designed to be easy to read, especially for anyone who is familiar with a language such as PL/1. Since it is intended for human readers rather than computers, it is somewhat informal at times, and we ignore the difficulty of implementing some of its features. The main features and conventions of the language are described in this section. The reader should use this language when writing the algorithms required by the exercises. Every algorithm begins with a description of its inputs and outputs, and the lines of an algorithm are numbered so they can be referred to in the text.

A *symbol* is a finite sequence of letters and digits; the first must be a letter. Symbols are used for variable and array names, for statement labels, and for names of fields in the nodes of data structures. Statement labels are followed by a colon and then the statement; for example:

$$\text{LOOP: } I \leftarrow I + 1$$

Arrays, variables, and fields of nodes may be of one of the following types: integer, real, character, logical, complex, bitstring, or pointer. Usually the type is clear from the context and declarations are not included; pointers, however, are almost always declared as such (in comments rather than in formal declarations).

Some instructions or subtasks in an algorithm will be stated in English; for example, "Let x be the largest entry in A," where A is an array, or "Insert x in L," where L is a linked list. This is done to avoid cluttering the algorithm with details not relevant to the main problem.

Arithmetic expressions may use the usual arithmetic operators $(+, -, *, /,$ and $\uparrow$ for exponentiation; $\uparrow$ and $*$ are often omitted.) Logical expressions may use the relational operators $=, \neq, <, >, \leq,$ and $\geq$, and the logical operators *and*, *or*, and *not*.

Assignment statements are statements of the form

$$a \leftarrow b$$

where a is a variable, array entry, or field of a node, and b is an arithmetic, logical, or pointer expression. The value of b is assigned to a. If a and b are both variables, array entries, or fields of nodes, the notation

$$a \leftrightarrow b$$

may be used to indicate that the contents of a and b are to be interchanged, and

$$a \leftarrow b \leftarrow c$$

may be used to indicate assignment of the value of an expression c to both a and b.

A **goto** statement has the form

$$\textbf{goto } label$$

It causes a branch to the statement with the specified label.

Conditional statements have either of the following forms:

$$\textbf{if } c \quad \textbf{then } s$$

or

$$\textbf{if } c \quad \textbf{then } s$$
$$\textbf{else } s'$$

where c is a logical expression and s and s' are single statements or sequences of statements enclosed between **do** and **end**. For both forms, if c is true, s is executed (once). If c is false, in the first form execution of the **if** statement is complete, and in the second s' is executed. In all cases, control then passes to the next statement unless a **goto** statement in s or s' caused a transfer of control elsewhere.

There are two loop instructions: **while** and **for**. The form of the **while** statement is

while *c* **do**

s

end

where *c* is a logical expression and *s* is a sequence of one or more statements. If *s* is a single, short statement, the following alternative form may be used:

while *c*

do *s* **end**

While *c* is true, *s* is executed. Before each execution of *s*, *c* is tested; if (when) *c* is false, control passes to the statement following the **while** statement. Note that if *c* is false when control first reaches the **while** statement, *s* is not executed at all. The effect of the **while** statement is equivalent to

LOOP: **if** *c* **then do**

s

goto LOOP

end

If the condition *c* in a **while** or **if** statement consists of several logical expressions connected by *and*'s and *or*'s, evaluation of the expressions is assumed to be left to right, with evaluation terminating as soon as the truth or falsity of *c* can be determined. Thus, for example, we assume an expression "$P \neq \Lambda$ and DATA$(P) \neq$ NAME" would not cause an illegal reference to DATA(Λ) when $P = \Lambda$.

The form of the **for** statement is

for *var* ← *init* **to** *limit* **by** *incr* **do**

s

end

where *var* is a variable, *init*, *limit*, and *incr* are arithmetic expressions, and *s* is a sequence of one or more statements. Initially *var* is assigned the value of *init*. If *incr* ≥ 0, then so long as *var* $\leq$ *limit*, *s* is executed and *incr* is added to *var*. (If *incr* < 0, *s* is executed and *incr* added to *var* so long as *var* $\geq$ *limit*.) The sign of *incr* may not be changed by *s*. The effect of the **for** statement, for *incr* ≥ 0, is equivalent to

var ← *init*

LOOP: **if** *var* $\leq$ *limit* **then do**

s

var ← *var* + *incr*

goto LOOP

end

Note that if *init* $>$ *limit*, *s* is not executed at all. If *incr* < 0, "**if** *var* $\leq$ *limit*" would be replaced by "**if** *var* $\geq$ *limit*" and *s* would not be executed at all if *init* $<$ *limit*. The "**by** *incr*" portion of the **for** statement is optional; if it is omitted, an increment of one is assumed.

An **exit** statement may be used to terminate execution of a **while** or **for** loop before the usual termination condition is met. **exit** causes a branch to the statement that follows the (innermost) **while** or **for** loop containing it.

return is used to indicate termination of execution of an algorithm; it is generally omitted if the algorithm terminates after the last instruction, and is used most often when undesirable conditions are detected. **return** may be followed by a message enclosed in apostrophes.

Comments within an algorithm are enclosed in square brackets [like this]. Various input/output instructions such as **read** and **output** are used occasionally.

The algorithm language relies mostly on ends of lines rather than semicolons or other punctuation to separate statements. Several short statements (e.g., assignment statements) may be written on one line separated by semicolons. Note that **do**'s and **end**'s are always paired.

1.3   *Analyzing Algorithms*

We analyze algorithms with the intention of improving them, if possible, and for choosing among several available for a problem. We will use the following criteria:

1.   Correctness

2.   Amount of work done

3.   Amount of space used

4.   Simplicity

5.   Optimality

We will discuss each of these criteria at length and give several examples of their application.

CORRECTNESS

There are three major steps involved in establishing the correctness of an algorithm. First, before we can even attempt to determine whether an algorithm is correct, we must have a clear understanding of what "correct" means. We may say that an algorithm is correct if when given a valid input it computes for a finite amount of time and produces the right answer. This definition is helpful if we know what the valid inputs are and what "the right answer" is. Thus showing that an algorithm is correct means making a precise statement about what result it is to produce when given certain specified inputs, and then proving the statement. There are two aspects to an algorithm: the problem solution method and the sequence of instructions for carrying it out. Establishing the correctness of the method and/or formulas used may require a long sequence of lemmas and theorems about the objects on which the algorithm works (for example, graphs, permutations, matrices). For example, the validity of the Gauss elimination method for solving systems of linear equations, depends on a number of theorems in linear algebra. Some of the methods used in algorithms in this book must be justified by such theorems.

Finally, we have the instructions themselves. If an algorithm is fairly short and straightforward, we generally use some very informal (and indescribable) means of convincing ourselves that the various parts do what we expect them to do. We may check some details carefully (e.g., initial and final values of loop counters), and hand-simulate the algorithm on a few small examples. None of this proves that it is correct, but informal techniques may suffice for small programs. Most programs written outside of classes are very large and very complex. To prove the correctness of a large program, we can try to break the program down into smaller segments, show that if all of the smaller segments do their jobs properly, then the whole program is correct, and then prove that each of the segments is correct. This task is made easier if (it may be more accurate to say: "This task is possible only if") algorithms and programs are written so that they can be broken down into disjoint segments that can be verified separately. This is one of the many strong arguments for structured programming. Most of the algorithms presented in this book are the small segments from which large programs are built, so we will not deal with the difficulties and complexities of proving the correctness of very long algorithms or programs.

One of the most useful techniques for rigorous proofs of correctness is mathematical induction. It is used to show that loops in an algorithm do what they are intended to do. For each loop we state conditions and relationships that we believe are satisfied by the variables and data structures used, and then verify that these conditions hold by inducting on the number of passes through the loop. The details of the proof require carefully following the instructions in the algorithm.

We will illustrate the use of induction by a very simple example: the sequential search algorithm for finding the location, or index, of a given item X in a list. The algorithm compares X to each list entry in turn until a match is found or the list is exhausted. If X is not in the list, the algorithm returns 0 as its answer.

Algorithm 1.1 SEQUENTIAL SEARCH

Input: L, n, X, where L is an array with n entries.

Output: j.

1. $j \leftarrow 1$
2.   **while** $j \leq n$ and $L(j) \neq X$
3.   **do** $j \leftarrow j + 1$ **end**
4.   **if** $j > n$ **then** $j \leftarrow 0$

Before trying to show that the algorithm is correct, we should make a precise statement about what it is intended to do. This seems simple enough; using the explanation above, we may state that if L is a list with n entries, the algorithm terminates with j set to the index of the list entry equal to X, if there is one, and set to 0 otherwise. This statement has two faults. It does not specify the result if X appears more than once in the list, and it does not indicate for which values of n the algorithm is expected to work. We may assume that n is nonnegative, but what if the list is empty, i.e., if $n = 0$? A more precise statement is:

Given an array L containing n items $(n \geqslant 0)$ and given X, the sequential search algorithm terminates with j set to the index of the first occurrence of X in L, if X is there, and set to 0 otherwise.

We will prove this statement by proving a stronger one which makes detailed claims about conditions that hold during execution of the algorithm. The stronger statement, in a form to which we can apply induction, is:

For $1 \leqslant k \leqslant n + 1$, if and when control reaches the tests in line 2 for the kth time, the following conditions are satisfied:

1. $j = k$ and for $1 \leqslant i < k, L(i) \neq X$.

2. If $k \leqslant n$ and $L(k) = X$, the algorithm will terminate with j still equal to k after executing the tests and line 4.

3. If $k = n + 1$, the algorithm will terminate with $j = 0$ after executing the tests and line 4.

Proof Let $k = 1$. Then $j = k$ from line 1; the second part of condition 1 is vacuously satisfied. For condition 2, if $1 \leqslant n$ and $L(1) = X$, the test in line 2 fails and control passes to line 4 where, since $j = k \leqslant n$, j is unchanged. For condition 3, if $k = n + 1$, then $j = n + 1$ so the test in line 2 fails and control passes to line 4 where j is set to zero. (Note that this is the case where $n = 0$ and the list is empty.)

Now we assume that the three conditions are satisfied for some $k < n + 1$ and show that they hold for $k + 1$. Suppose control has reached the tests in line 2 for the $(k + 1)$st time. Condition 1 for k and the fact that line 3 was executed once more before returning to line 2 for the $(k + 1)$st time imply that now $j = k + 1$. Condition 1 for k says that $L(i) \neq X$ for $1 \leqslant i < k$. Condition 2 for k implies that $L(k) \neq X$ since otherwise the algorithm would have terminated already; thus for $1 \leqslant i < k + 1$, $L(i) \neq X$ and condition 1 is established for $k + 1$. The arguments to establish conditions 2 and 3 for $k + 1$ are very similar to the arguments used when $k = 1$, so we omit them.   □

The claims we have proved show that the tests in line 2 are executed at most $n + 1$ times, that the output is $j = 0$ if and only if they were executed $n + 1$ times, and in that case for $1 \leqslant i < n + 1$, $L(i) \neq X$; i.e., X is not in L. The output is $j = k$ if and only if $L(k) = X$ and for $1 \leqslant i < k, L(i) \neq X$ so k is the index of the first occurrence of X in the array. Thus the algorithm is correct.

(Note that if we did not require the index of the *first* occurrence of X in L, but merely *any* occurrence, the algorithm could be shortened by searching the array backwards, starting with $j \leftarrow n$, and omitting line 4.)

If this proof seems a bit tedious, imagine what a proof of correctness of a full-sized program with complex data and control structures would be like. But if one wants to rigorously verify that a program is correct, this is the sort of work that must be done. Programmers rarely, if ever, write out such proofs in complete detail, but they should go through similar arguments to convince themselves that an algorithm or program works.

We will not do formal proofs of correctness in this book, though we will give arguments or explanations to justify complex or tricky parts of algorithms. The sample proof was presented here to give the reader a glimpse of what such proofs are like, and more basically, to show the reader that correctness *can be proved*, though indeed for long and complex (i.e., real) programs it is a formidable task.

AMOUNT OF WORK DONE

How shall we measure the amount of work done by an algorithm? The measure we choose should aid in comparing two algorithms for the same problem so that we can determine whether one is more efficient than the other. It would be handy if our measure of work gave some indication of how the actual execution times of the two algorithms compare, but we will not use execution time as a measure of work for a number of reasons. First, of course, it varies with the computer used, and we don't want to develop a theory for one particular computer. We may instead count the number of instructions or statements executed by a program, but this measure still has several of the other faults of execution time. It is highly dependent on the programming language used and on the programmer's style. It also would require that we spend time and effort writing and debugging programs for each algorithm to be studied. We want a measure of work that tells us something about the efficiency of the *method* used by an algorithm independent of not only the computer, programming language, and programmer, but also of the many implementation details or "bookkeeping" operations such as incrementing loop indexes, computing array indexes, or setting pointers in data structures. Our measure of work should be both precise enough and general enough to develop a rich theory that is useful for many algorithms and applications.

A simple algorithm may consist of some initialization instructions and a loop. The number of passes made through the loop to compute an answer is a fairly good indication of the work done by such an algorithm. However, there are some problems with this measure. The amount of work done in one pass through the loop may be much more than the amount done in another pass (and may be very hard to count). One algorithm may do many more operations in one pass through its loop than another algorithm does. Two algorithms for the same problem may have entirely different control structures; one may have one simple loop while the other may have several including nested loops.

In many cases, to analyze an algorithm we can isolate a particular operation fundamental to the problem under study (or to the types of algorithms being considered), ignore the bookkeeping, and count the number of chosen, or basic, operations performed by the algorithm.

Here are some examples of reasonable choices of basic operations for several problems.

Problem	Operation
1. Find X in a list of names.	Comparison of X with an entry in the list.
2. Multiply two matrices with real entries.	Multiplication of two real numbers (or multiplication and addition of real numbers).
3. Sort a list of numbers.	Comparison of two list entries.
4. Traverse a binary tree (represented as a linked structure where each node contains pointers to its left and right children).	Traversing a link. (Here, setting a pointer would be considered a basic operation rather than bookkeeping.)

We may choose a basic operation for a problem and then find that some algorithms do a substantial amount of other work that would not reasonably be considered book-keeping, it may be appropriate to count certain other operations as well. For example, we may have chosen multiplication as the basic operation for algorithms to compute a certain function and then discovered that some algorithms do very few multiplications but many additions. It would be reasonable to redefine the measure of work as both the number of multiplications and the number of additions performed. Since multiplication usually requires more time than addition, we may improve an algorithm by doing more of the latter and fewer of the former. Thus if we simply count the total number of basic operations performed, we would lose useful information about the relative merits of two algorithms. When there is more than one basic operation, they will be counted separately. We may assign weights to the totals to reflect the relative time required to perform them.

So long as the basic operation(s) is chosen well and the total number of operations performed is roughly proportional to the number of basic operations, we have a good measure of the work done by an algorithm and a good criterion for comparing several algorithms. The reader may not find this statement convincing now; we will be adding more justification for it throughout much of the discussion in the remainder of this section. For now, we simply point out that our definition of a measure of work allows a great deal of flexibility. Though we will often try to choose one, or at most two, basic operations, one can include as basic operations some of those that we call bookkeeping, and, in the extreme, one could choose as the basic operations the set of machine instructions for a particular computer. Thus by varying the choice of basic operations, we can vary the degree of precision and abstraction (i.e., imple-mentation—independence) of our analysis to fit our needs.

What if the total number of operations performed by an algorithm is not pro-portional to the number of basic operations? What if it is substantially higher? In the extreme case, we might choose a basic operation for a certain problem and then discover that some algorithms for the problem use such different methods that they do not do *any* of the operations to be counted. In such a situation, we would restrict our study to a particular *class of algorithms*, one for which the chosen measure of work is reasonable. Algorithms using other techniques for which a different choice of basic operation is appropriate could be studied separately. A class of algorithms for a problem is usually defined by specifying the operations that may be performed on the data. (The degree of formality of the specifications will vary; usually informal descriptions suffice.) Comparisons of two algorithms from different classes would not be very precise.

Throughout the preceding discussion, we have often used the phrase "the amount of work done by an algorithm." It could be replaced by the formal (and shorter) term "the complexity of an algorithm." *Complexity* means the amount of work done, measured by some specified *complexity measure*, which in our examples has been the number of specified basic operations performed. Note that, in this sense, complexity has nothing to do with how complicated or tricky an algorithm is; a very complicated algorithm may have low complexity. We will use the terms "complexity," "amount of work done," and "number of basic operations done" almost interchangeably in this book.

AVERAGE AND WORST-CASE ANALYSIS

Now that we have a general approach to analyzing the amount of work done by an algorithm, we need a way of presenting the results of the analysis concisely. The amount of work done cannot be described by a single number because the number of basic operations performed is not the same for all inputs. We observe first that the amount of work done usually depends on the size of the input. For example, alphabetizing a list of 1000 names usually requires more operations than alphabetizing a list of 100 names, using the same algorithm. Solving a system of twelve linear equations in twelve unknowns generally takes more work than solving a system of two linear equations in two unknowns. The second important observation is that even if we consider only inputs of one size, the number of operations performed by an algorithm may depend on the particular input. An algorithm for alphabetizing a list of names may do very little work if only a few of the names are out of order, but it may have to do much more work on a list that is very scrambled. Solving a system of twelve linear equations may not require much work if most of the coefficients are zero.

The first observation indicates that we need a measure of the size of the input for a problem. It is usually easy to choose a reasonable measure of size. Some examples are given below.

Problem	Size of input
1. Find *X* in a list of names.	The number of names in the list.
2. Multiply two matrices.	The dimensions of the matrices.
3. Sort a list of numbers.	The number of entries in the list.
4. Traverse a binary tree.	The number of nodes in the tree.
5. Solve a system of linear equations.	The number of equations, or the number of unknowns, or both.
6. Solve a problem concerning a graph.	The number of nodes in the graph, or the number of edges, or both.

The second observation we made above was that even if the input size is fixed at, say, *n* the number of operations performed may depend on the particular input. How, then, are the results of the analysis of an algorithm to be expressed? It is not desirable (and not always possible) to list the number of operations performed by a particular algorithm on each input of size *n*. A possible solution is to compute the *average behavior* of the algorithm; that is, compute the number of operations performed for each input of size *n* and then take the average. This result may not be particularly useful in practice because some inputs may occur much more frequently than others. A weighted average will give a more meaningful result. Let D_n be the set of inputs of size *n* for the problem under consideration. Let *I* be an element of D_n and let $p(I)$ be the probability that input *I* occurs. Let $t(I)$ be the number of basic operations performed by the algorithm of input *I*. Then its average behavior may be realistically defined as

$$A(n) = \sum_{I \in D_n} p(I) \cdot t(I);$$

$t(I)$ is to be computed by careful examination and analysis of the algorithm, but $p(I)$ cannot be computed analytically. The function p must be determined from experience and/or special information about the application for which the algorithm is to be used; it is usually not easy to determine. If p is complicated, the computation of average behavior is difficult. Also, of course, if p depends on a particular application of the algorithm, the function A describes the average behavior of the algorithm for only that application.

Another approach to describing the behavior of an algorithm is to compute its *worst-case complexity*. Using the notation of the previous paragraph, we define the function W by

$$W(n) = \max_{I \in D_n} t(I).$$

$W(n)$ is the maximum number of basic operations performed by the algorithm on any input of size *n*. It can usually be computed more readily than $A(n)$ can. It is valuable because it gives an upper bound on the work done by the algorithm. The worst-case analysis could be used to help form an estimate for a time limit for a particular implementation of an algorithm. This is particularly useful in real-time applications. We will do worst-case analysis for most of the algorithms presented in this book and, unless otherwise stated, whenever we refer to the amount of work done by an algorithm, we will mean the amount of work done in the worst case.

The following examples illustrate average- and worst-case analysis.

Example 1.1

Problem: Let *L* be an array containing *n* entries. Find an index of a specified entry *X*, if *X* is in the list; return 0 as the answer if *X* is not in the list.

Algorithm: See Algorithm 1.1.

Basic operation: Comparison of *X* with a list entry.

Average-behavior analysis: The inputs for this problem can be categorized according to where in the list *X* appears, if it appears at all. That is, there are $n + 1$ inputs to consider. For $1 \leqslant i \leqslant n$, let I_i represent the case where *X* is in the *i*th position in the list and let I_{n+1} represent the case where X is not in the list at all. Then, let $t(I)$ be the number of comparisons done (the number of times the condition $L(j) \neq X$ in line 2 is tested) by the algorithm on input I. Obviously, for $1 \leq i \leq n$, $t(I_i) = i$, and $t(I_{n+1}) = n$. To compute the number of comparisons done on the average, we must know how likely it is that X is in the list and how likely it is that X is in any particular position. Let q be the probability that X is in the list, and assume that each position is equally likely. Then for $1 \leq i \leq n$, $p(I_i) = q/n$ and $p(I_{n+1}) = 1 - q$. Thus,

$$A(n) = \sum_{i=1}^{n+1} p(I_i)t(I_i)$$

$$= \sum_{i=1}^{n} \frac{q}{n} i + (1 - q)n$$

$$= \frac{q}{n} \sum_{i=1}^{n} i + (1 - q)n$$

$$= \frac{q}{n} \frac{n(n + 1)}{2} + (1 - q)n$$

$$= q \frac{(n + 1)}{2} + (1 - q)n.$$

If, for example, it is known that X is in the list, then $q = 1$ and $A(n) = (n + 1)/2$, which says that on the average, half the list must be examined. If $q = \frac{1}{2}$, that is, if there is a 50–50 chance that X is not in the list, then $A(n) = [(n + 1)/4] + n/2 \approx \frac{3}{4}n$. Roughly $\frac{3}{4}$ of the entries are examined, on the average, in this case.

Worst-case analysis: Clearly $W(n) = \max\{t(I_i): 1 \leq i \leq n + 1\} = n$. The worst cases occur when X is the last entry in the list and when X is not in the list at all. In both of these cases X is compared to all n entries.

Example 1.1 illustrates how we should interpret D_n, the set of inputs of size n. Rather than consider all possible lists of names, numbers, or whatever, that could occur as inputs, we identify the properties of the inputs that affect the behavior of the algorithm; in this case whether X is in the list at all, and if so, where it is. An element I in D_n may be thought of as a set (or equivalence class) of all lists and values for X such that X occurs in the specified place in the list (or not at all). Then $t(I)$ is the number of operations done for any one of the inputs in I.

Observe also that the input for which an algorithm behaves worst depends on the particular algorithm, not on the problem. For Algorithm 1.1 a worst case occurs when the only position in the list containing X is the last. For an algorithm that searched the list from the bottom up, a worst case would occur if X were only in the first position. (Another worst case would again be when X is not in the list at all.)

Example 1.2

Problem: Let $A = (a_{ij})$ and $B = (b_{ij})$ be two $n \times n$ matrices with real entries. Compute the product matrix $C = AB$.

Algorithm: Use the algorithm implied by the definition of the matrix product:

$$c_{ij} = \sum_{k=1}^{n} a_{ik}b_{kj} \qquad \text{for } 1 \leqslant i, j \leqslant n.$$

Algorithm 1.2 MATRIX MULTIPLICATION

> **for** $i \leftarrow 1$ **to** n **do**
> > **for** $j \leftarrow 1$ **to** n **do**
> > > $c_{ij} \leftarrow 0$
> > > **for** $k \leftarrow 1$ **to** n **do** $c_{ij} \leftarrow c_{ij} + a_{ik}b_{kj}$ **end**
> >
> > **end**
>
> **end**

Basic operation: Multiplication of matrix entries.

Analysis: For each entry of C, n multiplications are done. C has n^2 entries so

$$A(n) = W(n) = n^3.$$

Example 1.2 illustrates that for some algorithms the instructions performed, hence the amount of work done, are independent of the input. Thus if n is the input size, $A(n) = W(n) =$ the number of basic operations done in all cases. In other algorithms for the same problem, this may not be true.

The concepts of worst-case and average-behavior analysis would be useful even if we had chosen a different measure of work done (including those briefly mentioned earlier: execution time and number of passes through a loop). The observation that the amount of work done often depends on the size and properties of the input would lead to the discussion of average behavior and worst-case behavior, no matter what measures were used.

We return now to our discussion of the merits of our measure of work. Just how useful is it to count one operation and ignore the rest? We have said that the choice of a basic operation is acceptable only if the total number of operations performed is roughly proportional to the number of basic operations. More precisely, we mean that if an algorithm does $t(I)$ basic operations for an input I, the total number of operations would be at most $ct(I)$ and the actual execution time $c't(I)$ seconds, where c and c' are constants; they depend on the algorithm and the computer on which it is implemented, but not on I. (The previous statement may not hold for some inputs; there may be a few 'special cases' in which no basic operations are needed, but of course the algorithm does more than zero operations.) We can express a similar idea in terms of W and A, which describe the worst-case and average complexity of an algorithm as functions of input size: If an algorithm does $W(n)$ ($A(n)$) basic operations in the worst case (on the average), it does at most $cW(n)$ ($cA(n)$) operations in total and runs in at most $c'W(n)$ ($c'A(n)$) seconds in the worst case (on the average). (Here also, we should allow for some special cases; the previous statement may not hold for very small values of n. For example, for $n = 0$, the sequential search algorithm does no comparisons of X with list entries, but it does *some* work — it must determine that $n = 0$ and output an answer.)

If it were the case that most of the algorithms people developed for a given problem did approximately the same amount of work, then to compare algorithms we would need a very precise count of all the work done including bookkeeping. That is, it would be very important to discover not only W and A for an algorithm, but also the constant c in the discussion above. However, for a great many problems, some algorithms have been developed that are so much better than others that the value of c for each of them is not very important. For example, if for one algorithm $W_1(n) = n^2$ and for another $W_2(n) = \frac{5}{2}n$, the latter will run faster (in the worst case) for almost all input sizes even if it does more bookkeeping per basic operation than the former. Specifically, let $c_1 = 3$ and $c_2 = 10$; so the first algorithm does $3n^2$ operations in total and the second does $25n$ (in the worst case). Thus so long as $n > 8$ (which would probably be the case for most problems solved by computer), the second algorithm does less work. No matter what (positive) values are chosen for c_1 and c_2, $c_2 W_2(n)$ would be less than $c_1 W_1(n)$ for all large n, that is, for all n larger than some particular integer. The function W_2 is said to be of lower order than W_1. This notion of the order of a function is very useful and we will take some time now to discuss and illustrate it.

Let f and g be functions with domain **N**, the natural numbers. We say that the order of f is less than or equal to the order of g, denoted "f is $O(g)$", or "$f = O(g)$," if there are positive constants c and n_0 such that $f(n) \leqslant cg(n)$ for all $n > n_0$. Thus $\frac{5}{2}n$ is $O(n^2)$. Observe that $n^2/2$ is $O(307n^2)$ and $307n^2$ is $O(n^2/2)$ and, in general, two functions that differ by a constant factor are of the same order. The set of functions of the same order as f, $\Theta(f)$, includes all functions g such that f is $O(g)$ and g is $O(f)$. Although $\Theta(f)$ is defined as a set, it is conventional to write "$g = \Theta(f)$" or "g is $\Theta(f)$," rather than "$g \in \Theta(f)$." Thus $307n^2 = \Theta(n^2/2)$. Low-order terms in a function do not affect its order. Thus $n(n-1)/2 = \Theta(n^2)$ and $5n^4 + 2n^3 - 7 = \Theta(n^4)$. Note that n^4 is not $O(n^2)$. (The reader should verify these statements.) We usually take a function with a simple formula as a canonical representative of its order, e.g., n^2 and n^4 in the above examples. Thus when we speak of finding the order of a function, we usually mean finding a simpler function of the same order.

There are several useful techniques for showing that a function is or is not of the same or lower order than another. Proofs by contradiction often work for the negative statements. For example, to show that n^3 is not $10(n^2)$, assume first that it is. Then for some constants c and n_0, $n^3 \leqslant cn^2$ for $n \geqslant n_0$. Thus $n \leqslant c$ for all $n \geqslant n_0$, but this is impossible. Another technique comes from an alternate description of $\Theta(f)$ using limits. If $\lim_{n \to \infty} f(n)/g(n) = c$ for some constant $c \neq 0$, then f and g are of the same order. If $c = 0$, $f = O(g)$, but g is not $O(f)$; i.e., f is of lower order than g. And if $c = \infty$, then f is of higher order than g. [For example, $\lim_{n \to \infty} (n^2/2)/307n^2 = 1/614$.] L'Hôpital's Rule is often useful for computing the limit. It states that if

$$\lim_{n \to \infty} f(n) = \lim_{n \to \infty} g(n) = \infty,$$

then

$$\lim_{n \to \infty} \frac{f(n)}{g(n)} = \lim_{n \to \infty} \frac{f'(n)}{g'(n)},$$

where f' and g' are the derivatives of f and g. It can be used when these limits and derivatives exist. Thus to see that $\log n$ is $O(n)$ but not $\Theta(n)$, we compute

$$\lim_{n \to \infty} \frac{\log n}{n} = \lim_{n \to \infty} \frac{\ln n \log e}{n} = \lim_{n \to \infty} \frac{\frac{1}{n} \log e}{1}$$

$$= \lim_{n \to \infty} \frac{1}{n} \log e = 0.$$

The reader should do Exercises 1.1 and 1.2 to gain some facility with the ideas and techniques presented in the previous two paragraphs.

Now, when we analyze a new algorithm the first thing we want to know is whether the amount of work done is of lower (or the same, or higher) order than the amount done by other algorithms already studied. For many problems, new algorithms have been found with complexity of lower order than algorithms already known and used. Since the total execution time is of the same order as the number of basic operations done, we are justified in counting basic operations and ignoring bookkeeping and implementation details! Of course, when the order of the amount of work done is the same for two algorithms, then we may compare the exact number of basic operations and the number of bookkeeping operations per basic operation. The same approach should guide the development and improvement of one algorithm: First look for modifications that reduce the order of the work done; then worry about details that reduce the work without affecting the order. The latter may be thought of as fine tuning.

SPACE USAGE

The number of memory cells used by a program, like the number of seconds required to execute a program, depends on the particular implementation. However, some conclusions about space usage can be made by merely examining an algorithm. A program will require storage space for the instructions, the constants and variables used by the program, and the input data. It may also use some workspace for manipulating the data and storing information needed to carry out its computations. The input data itself may be representable in several forms, some of which require more space than others. If the input data has one natural form — for example, an array of numbers or a matrix — then we analyze the amount of *extra* space used, aside from the program and the input. If the amount of extra space is constant with respect to the input size, the algorithm is said to work *in place*. This term is used especially in reference to sorting algorithms. If the input can be represented in various forms (for example, graphs), then, of course, we will consider the space required for the input itself as well as any extra space used. In general, we will refer to the number of "cells" used without precisely defining cells. The reader may think of a cell as being large enough to hold one number. If the amount of space used depends on the particular input, worst-case and average-case analysis can be done.

SIMPLICITY

It is often, though not always, the case that the simplest and most straightforward way of solving a problem is not the most efficient. Yet simplicity in an algorithm is a desirable feature. It may make verifying the correctness of the algorithm easier, and it makes writing, debugging, and modifying a program for the algorithm easier. The time needed to produce a debugged program should be considered when choosing an algorithm, but if the program is to be used very often, its efficiency will probably be the determining factor in the choice.

OPTIMALITY

To analyze the complexity of a problem, as opposed to that of a specific algorithm, we choose a class of algorithms (often by specifying the types of operations the algorithms will be permitted to perform) and a measure of complexity, for example, the basic operation(s) to be counted. Then we may ask how many operations are actually *needed* to solve the problem. Is there a "best" possible algorithm? We say that an algorithm is *optimal* (in the worst case) if there is no algorithm in the class under study that performs fewer basic operations (in the worst case).

Does every algorithm in the class have to be individually analyzed before we can conclude that one is optimal? Fortunately, no; we can prove theorems that establish a lower bound on the number of operations needed to solve a problem. Then any algorithm that performs that number of operations would be optimal. Thus there are two tasks to be carried out in order to find a good algorithm in a class of algorithms being studied, or from another point of view, to answer the theoretical question: How much work is necessary and sufficient to solve the problem?

1.   Devise what seems to be an efficient algorithm; call it A. Analyze A and find a function W such that for inputs of size n, A does at most $W(n)$ basic operations.

2.   For some function F, prove a theorem stating that for any algorithm in the class under consideration, there is some input of size n for which the algorithm must perform at least $F(n)$ basic operations.

If the functions W and F are equal, then the algorithm A is optimal. If not, it may be that there is a better algorithm or that there is a better lower bound. Observe that analysis of a specific algorithm gives an *upper bound* on the number of basic operations necessary to solve a problem, and a theorem of the type described in (2) gives a *lower bound* on the number of operations necessary by any algorithm in the class under study, in the worst case.

The concept of a lower bound for the worst-case behavior of algorithms in a particular class is very important in computational complexity. Example 1.3 below and the problems studied in the following sections and in Chapter 2 will help to clarify the meaning of lower bounds and illustrate techniques for establishing them. The reader should keep in mind the definition: "F is a lower bound for a class of algorithms" means that for *any* algorithm in the class, and any input size n, there is *some* input of size n for which the algorithm must perform *at least* $F(n)$ basic operations.

Example 1.3

Problem: Find the largest entry in a list of n numbers.

Class of algorithms: Algorithms that can compare and copy list entries, but do no other operations on them.

Basic operation: Comparison of two list entries.

Upper bound: Suppose the numbers are in an array L. The following algorithm finds the maximum.

Algorithm 1.3 FINDMAX

Input: L, an array of numbers; $n \geqslant 1$, the number of entries.

Output: MAX, the largest entry in L.

1.   MAX $\leftarrow L(1)$; $i \leftarrow 2$
2.   **while** $i \leqslant n$ **do**
3.   **if** MAX $< L(i)$ **then** MAX $\leftarrow L(i)$
4. $i \leftarrow i + 1$
     end

Comparisons of list entries are done in line 3, which is executed exactly $n - 1$ times. Thus $n - 1$ is an upper bound on the number of comparisons necessary to find the maximum in the worst case. Is there an algorithm that does fewer?

Lower bound : To establish a lower bound we may assume that the entries in the list are all distinct. This assumption is permissible because if we can establish a lower bound on worst-case behavior for some subset of inputs (lists with distinct entries), it is a lower bound on worst-case behavior when all valid inputs are considered.

   In a list with n distinct entries, $n - 1$ entries are *not* the maximum. We can conclude that a particular entry is not the maximum only if it is smaller than at least one other entry in the list. Hence, $n - 1$ entries must be "losers" in comparisons done by the algorithm. Each comparison has only one loser, so at least $n - 1$ comparisons must be done. Thus $F(n) = n - 1$ is a lower bound on the number of comparisons needed.

Conclusion : Algorithm 1.3 is optimal.

   We could take a slightly different point of view to establish the lower bound in Example 1.3. If we are given an algorithm and a list of n numbers such that the algorithm halts and produces an answer after doing less than $n - 1$ comparisons, then we can prove that the algorithm gives the *wrong* answer for some set of input data. For if no more than $n - 2$ comparisons are done, two entries are never losers, that is, they are not known to be smaller than any other entries. The algorithm can specify at most one of them as the maximum. We can simply replace the other with a higher number (if necessary). Since the results of all comparisons done will be the same as before, the algorithm will give the same answer and it will be wrong. This argument is a proof by contradiction. It illustrates a useful technique for establishing lower bounds, namely, to show that if an algorithm does not do enough work, one can arrange the input so that the algorithm gives the wrong answer.

Example 1.4

Problem: Let $A = (a_{ij})$ and $B = (b_{ij})$ be two $n \times n$ matrices with real entries. Compute the product matrix $C = AB$. (This problem will be discussed much more thoroughly in Chapter 5.)

Class of algorithms: Algorithms that can perform multiplications, divisions, additions, and subtractions on the matrix entries and on the intermediate results obtained by performing these operations on the entries.

Basic operation: Multiplication.

Upper bound: The usual algorithm (see Example 1.2) does n^3 multiplications; hence at most n^3 multiplications are necessary.

Lower bound: It has been proved that at least n^2 multiplications are necessary.

Conclusions: There is no way to tell from the information given whether or not the usual algorithm is optimal. Some researchers have been trying to improve the lower bound, that is, to prove that more than n^2 multiplications are necessary, while others have looked for better algorithms. To date it has been shown that the usual algorithm is *not* optimal; there is a method that requires approximately $n^{2.8}$ multiplications. Is this method optimal? The answer is not known. The lower bound has not yet been improved.

In practice it may not be easy to show that an algorithm is optimal. For many problems, if F is a lower bound for a problem and W describes the worst-case behavior of a particular algorithm, we will be content with the algorithm if $W = \Theta(F)$. No algorithm can do the job for large n with substantially fewer basic operations.

Up to now we have been discussing optimality of worst-case behavior. What about optimality of average behavior? We can use the same approach that we use with worst-case behavior. Choose what seems to be a good algorithm and figure out the function A such that the algorithm does $A(n)$ basic operations, on the average, for inputs of size n. Then prove a theorem saying that any algorithm in the class being studied must perform $G(n)$ basic operations on the average for inputs of size n. If $A = \Theta(G)$, the average behavior of the algorithm is of optimal order. If not, look for a better algorithm or a better lower bound (or both).

We can use the same scheme to investigate optimality of space usage. Analyze a particular algorithm to get an upper bound on the amount of space needed, and prove a theorem to establish a lower bound. Can we find one algorithm for a given problem that is optimal with respect to both the amount of work done and the amount of space used? The answer to this question is: sometimes. Sometimes there is a trade-off between the two.

IMPLEMENTATION AND PROGRAMMING

Implementation is the task of turning an algorithm into a computer program. Since the ways in which algorithms are described vary from detailed sets of computer language-like instructions for manipulating variables and data structures to very abstract, high-level descriptions in English of solution methods for abstract problems, making no mention of computer representations of the objects involved, the implementation of an algorithm may be a fairly straightforward translating job or it may be a very lengthy and difficult job requiring a number of important decisions on the part of the programmer, particularly concerning the choice of data structures. Where appropriate, we will discuss implementation in the general sense of choosing data structures and describing ways to carry out instructions given in an English description of an algorithm. Such discussion is included for two reasons. One, it is a natural and important part of the process of producing a (good) working program. Two, consideration of implementation details is often necessary for analyzing an algorithm; the amount of time required for performing various operations on abstract objects such as sets and graphs depends on how these objects are represented. For example, forming the union of two sets may require only one or two operations if the sets are represented as linked lists, but would require a large number of operations, proportional to the number of elements in one of the sets, if they are represented as arrays and one must be copied into the other.

In the narrow sense, implementation, or simply programming, means converting a fairly detailed description of an algorithm and the data structures it uses into a program for a particular computer. Our analysis will be implementation-independent in this sense; in other words, it will be independent of the computer and programming language used and of many minor details of the algorithm or program.

A programmer can refine the analysis of algorithms under consideration using information about the particular computer to be used. For example, if more than one operation is counted, the operations can be weighted according to their execution times; or if the programmer has a choice of several algorithms, estimates of the actual number of seconds each would require (in the worst or average case) can be made. Sometimes knowledge of the computer used will lead to a new analysis. For example, if the computer has any unusual, powerful instructions that can be used effectively in the problem at hand, then one can study the class of algorithms that make use of those instructions and count them as the basic operations. If the computer has a very limited instruction set that makes implementation of the basic operation awkward, a different class of algorithms may be considered. Generally, however, if the implementation-independent analysis has been done well (with a reasonable choice of basic operation), then the program-dependent analysis should serve mainly just to add a bit more detail.

A detailed analysis of the amount of space used by the algorithms being studied is, of course, also appropriate when particular implementations are being considered.

Any special knowledge about the inputs to the problem for which an algorithm is sought can be used to refine the analysis. If, for example, the inputs will be restricted to a certain subset of all possible inputs, a worst-case analysis can be done for that subset. As we have noted, a good average-behavior analysis depends on knowing the probability of the various inputs occurring.

1.4   *Searching an Ordered List*

To illustrate the ideas presented in the previous section, we will carefully study a problem for which the reader has probably already seen and possibly programmed some algorithms.

The problem : Given an array L containing n entries sorted in nondecreasing order and given a value X, find an index of X in the list or, if X is not in the list, return 0 as the answer.

Observe that the sequential search algorithm (Algorithm 1.1) solves this problem, but it makes no use of the fact that the entries in the list are in order. Can we modify the algorithm so that it uses the added information and does less work? The first improvement is prompted by the observation that since the array is in nondecreasing order, as soon as an entry larger than X is reached, the algorithm can terminate with the answer 0. How does this change affect the analysis? Clearly, the modified algorithm is better in some cases; that is, it will terminate sooner for some inputs. The worst-case complexity, however, remains unchanged. If X is the last entry in the list or if X is larger than all the entries, then the algorithm will do n comparisons. Is the modified algorithm better on the average? We saw that if there is a 50–50 chance that X is in the list, and if all positions in the list are equally likely to be the location of X, then, on the average, the sequential search algorithm does about $\frac{3}{4}n$ comparisons. For the modified algorithm, we also must know how likely it is that X is *between* any two list entries. Suppose we define a *gap*, g_i, to be the set of values y such that $L(i-1) < y < L(i)$ for $i = 2, \ldots, n$. Also, let g_1 be all values less than $L(1)$ and g_{n+1} all values greater than $L(n)$. Now, to do an average-behavior analysis of the modified algorithm, let us assume that there is a 50–50 chance that X is in the list, that if it is, all positions in the list are equally likely (so have probability $1/2n$), and that if X is not in the list all gaps are equally likely (i.e., have probability $1/2(n + 1)$). For $1 \leq i \leq n$, it takes i comparisons to determine that $X = L(i)$ or that X is in g_i, and it takes n comparisons to determine that X is in g_{n+1}. So we compute the average number of comparisons as follows:

$$A(n) = \sum_{i=1}^{n} \frac{1}{2n} i + \sum_{i=1}^{n} \frac{1}{2(n + 1)} i + \frac{1}{2(n + 1)} n.$$

The first term corresponds to cases in which X is in the list, and the latter two terms correspond to cases in which X is not in the list. Evaluating the sums is easy and left as an exercise. The result is that $A(n)$ is roughly $n/2$.

We have not made a substantial improvement over Algorithm 1.1, but the effort was not wasted; we do not expect to think up the ideal algorithm for every problem on the first attempt, and the more easy examples of algorithm analysis that the reader sees, the more prepared he or she will be to tackle complicated problems.

Now let us try again. Can we find X in an ordered list of n entries by some method that does substantially less than n comparisons in the worst case? Suppose we write an algorithm that compares X to, say, every fourth entry in the list. If there is a match, the algorithm terminates. If at some point the entry to which X is compared is larger than X, then X lies between the last two entries to which it was compared. A few more comparisons (how many?) will suffice to determine the position of X if it is in the list or to determine that it is not there. The details of the algorithm and the analysis are left for the reader but it is easy to see that only about one fourth of the entries in the table are examined, that is, are compared to X. Thus in the worst case approximately $n/4$ comparisons are done. This is certainly an improvement over the sequential search algorithm, but can we do even better?

We could pursue the same scheme, choosing large values for k and designing an algorithm that compares X to every kth entry, but while we would be lowering the number of comparisons needed to find a subsection of the list that contains X (if it is in the list at all), we would also be increasing the size of the subsection and hence the number of additional comparisons needed to locate X within it.

Suppose we adopt a slightly different strategy. Instead of choosing a particular integer k and comparing X to every kth entry, we could compare X first to the entry in the middle of the list. If X is larger, then, if it is in the list at all, it is in the second half; with one comparison the entire first half of the list can be eliminated from consideration. Conversely, if X is smaller than the entry in the middle of the list, the second half of the list may be eliminated from consideration. (Of course, if X is equal to the middle entry, there is nothing more to do.) Until X is found or it is determined that X is not in the list, compare X to the middle entry in the section of the list under consideration. After each comparison, the size of the section of the list that may contain X is cut in half. This method is called binary search. The binary search algorithm is given below.

Algorithm 1.4 BINARY SEARCH

Input: $L, n \geqslant 0$, and X, where L is an array with n entries and X is the item sought.

Output: j such that $L(j) = X$ if X is in L and $j = 0$ if X is not in L.

1. $k \leftarrow 1; m \leftarrow n$ [k and m are the indexes of the first and last entries, respectively, of the section of the array currently being searched.]
2.   while $k \leqslant m$ do
3. $j \leftarrow \left\lfloor \dfrac{k+m}{2} \right\rfloor$ [index of the middle entry]
4.   if $X = L(j)$ then return
5.   if $X < L(j)$ then $m \leftarrow j - 1$
6.   else $k \leftarrow j + 1$
     end
7. $j \leftarrow 0$

WORST-CASE ANALYSIS

Let $W(n)$ be the number of basic operations (that is, comparisons of X to list entries) performed by the binary search algorithm in the worst case on lists with n entries. We will assume that one comparison with a three-way branch is done for the tests on X in lines 4 and 5; thus $W(n)$ is the number of passes through the **while** loop. Suppose $n > 1$. The first time line 2 is encountered, the task of the algorithm is to find X in a list of n entries indexed from $k = 1$ to $m = n$. It proceeds to line 4 and compares X to $L(\lfloor (1 + n)/2 \rfloor)$. In the worst case these are unequal and either k or m will be changed so that on the next pass through the loop, the task is to find X in the section of the list indexed from k to m, inclusive. How many entries are there in this section? If n is even, there are $n/2$ entries in the section of the list following $L(\lfloor (1 + n)/2 \rfloor)$ and $(n/2) - 1$ entries in the section preceding it. If n is odd there are $(n - 1)/2$ entries in both sections. Hence, there are at most $\lfloor n/2 \rfloor$ entries in the section of the list in which the algorithm will look for X on the next pass through the loop. The number of comparisons done by the algorithm beginning with the second pass through the loop is the same as the number it would do if the input were a list with $\lfloor n/2 \rfloor$ entries. Thus $W(n) = 1 + W(\lfloor n/2 \rfloor)$. This is an example of a *recurrence relation*. A recurrence relation for a function is an equation relating the value of the function on the argument n to values of the function on smaller arguments. They occur often in the analysis of algorithms. The recurrence relation alone does not provide enough information to determine W, but W could be evaluated for arbitrary n if its value on the smallest argument in its domain were known. It is easy to see from the algorithm that $W(0) = 0$. We will also compute $W(1)$.

If $n = 1$, then $k = m = j = 1$ and X is compared to $L(1)$. If the **return** in line 4 is not executed, either m will be set to 0 or k to 2, and the instructions in the loop (in particular, the comparison) are not executed again. Thus, whether or not X is in the list, one comparison is done, so $W(1) = 1$. Thus what is known about W can be expressed in two equations:

$$W(n) = 1 + W\left(\left\lfloor \frac{n}{2} \right\rfloor\right) \qquad \text{for } n > 1$$

$$W(1) = 1.$$

When a function is described by a recurrence relation, an equation giving the value of the function for a particular argument is called a *boundary condition*. The equation $W(1) = 1$ is a boundary condition for W.

Expanding the recurrence relation a few times should enable us to make a good guess at a formula describing $W(n)$ as a function of n without referring to other values of W. Such a formula is called a *closed form*. Expanding the recurrence relation gives

$$W(n) = 1 + W\left(\left\lfloor \frac{n}{2} \right\rfloor\right)$$

$$= 1 + 1 + W\left(\left\lfloor \frac{n}{2^2} \right\rfloor\right)$$

$$= 1 + 1 + 1 + W\left(\left\lfloor \frac{n}{2^3} \right\rfloor\right).$$

Each time the argument is divided by 2, 1 is added to the value of the function. Thus there will be approximately $\log n$ terms in the sum, each equal to 1; but $\log n$ may not be an integer so a slight adjustment is necessary. The exact formula is established by induction on n.

Lemma 1.1 $W(n) = \lfloor \log n \rfloor + 1$, for $n \geq 1$.

Proof, by induction on n. For the basis of the induction, let $n = 1$. Then we see that $\lfloor \log n \rfloor + 1 = \lfloor \log 1 \rfloor + 1 = 0 + 1 = 1 = W(1)$, by the boundary condition. For the induction step, assume that $n > 1$ and that for $1 \leq k < n$, $W(k) = \lfloor \log k \rfloor + 1$. Then $W(n) = $ (by the recurrence relation) $1 + W(\lfloor n/2 \rfloor) = $ (by the induction hypothesis)

$$1 + \left\lfloor \log \left\lfloor \frac{n}{2} \right\rfloor \right\rfloor + 1 = 2 + \left\lfloor \log \left\lfloor \frac{n}{2} \right\rfloor \right\rfloor$$

$$= \begin{cases} 2 + \lfloor \log n - 1 \rfloor & \text{if } n \text{ is even since } \left\lfloor \frac{n}{2} \right\rfloor = \frac{n}{2} \\ 2 + \lfloor \log(n-1) - 1 \rfloor & \text{if } n \text{ is odd since } \left\lfloor \frac{n}{2} \right\rfloor = \frac{n-1}{2} \end{cases}$$

$$= \begin{cases} 1 + \lfloor \log n \rfloor & \text{if } n \text{ is even} \\ 1 + \lfloor \log(n-1) \rfloor & \text{if } n \text{ is odd}. \end{cases}$$

If n is odd $\lfloor \log n \rfloor = \lfloor \log(n-1) \rfloor$, so in all cases $W(n) = 1 + \lfloor \log n \rfloor$. □

Theorem 1.2 The number of comparisons of X with list entries done by the binary search algorithm, in the worst case, for a list of n entries, is $\lfloor \log n \rfloor + 1$ for $n \geq 1$.

Observe that the binary search does fewer comparisons in the worst case than a sequential search does on the average. Also observe that the total number of operations performed by the binary search algorithm is (roughly) proportional to $\lfloor \log n \rfloor$ because one comparison is done in each pass through the **while** loop.

AVERAGE-BEHAVIOR ANALYSIS

For $1 \leq i \leq n$, let I_i represent all inputs for which the output of Algorithm 1.4 is $j = i$ (i.e., $X = L(i)$). For $2 \leq i \leq n$, let I_{n+i} represent inputs for which $L(i-1) < X < L(i)$. I_{n+1} and I_{2n+1} represent inputs where $X < L(1)$ and $X > L(n)$, respectively. Thus there are $2n + 1$ positions that X may occupy: the n positions in L and the $n + 1$ gaps. Let $t(I_i)$ be the number of comparisons of X with list entries done by Algorithm 1.4 on input I_i. Table 1.1 shows the values of t for $n = 25$. Observe that most inputs are worst cases; that is, it takes five comparisons to find X most of the time. So if we assume that all positions (including gaps) are equally likely, it is not unreasonable to expect the number of comparisons needed on the average to be close to $\log n$. Computation of the average yields $\frac{223}{51}$, or approximately 4.37, and $\log 25 \approx 4.65$.

Table 1.1

The number of comparisons done by binary search depending on the location of X; $n = 25$.

i	$t(I_i)$	i	$t(I_i)$
1	4	14	4
2	5	15	5
3	3	16	3
4	4	17	4
5	5	18	5
6	2	19	2
7	4	20	4
8	5	21	5
9	3	22	3
10	5	23	5
11	4	24	4
12	5	25	5
13	1	gaps { 26, 29, 32, 39, 42, 45	4
		all others	5

We will derive an approximate formula for the number of comparisons done on the average, given two assumptions:

1.   All positions (including gaps) are equally likely, so for $1 \leqslant i \leqslant 2n + 1$, $p(I_i) = 1/(2n + 1)$.

2. $n = 2^k - 1$, for some integer $k \geqslant 1$.

The second assumption is made to simplify the analysis a little. The result for all values of n is very close to the result that we will obtain.

Observe that $k = \lfloor \log n \rfloor + 1$, the number of comparisons done in the worst case. For $1 \leqslant t \leqslant k$, let s_t be the number of inputs for which the algorithm does t comparisons. For example, for $n = 25$, $s_3 = 4$ because three comparisons would be done for each of the four inputs I_3, I_9, I_{16}, and I_{22}. It is easy to see that $s_1 = 1 = 2^0$, $s_2 = 2 = 2^1$, $s_3 = 4 = 2^2$, and in general for $t < k$, $s_t = 2^{t-1}$. The algorithm does k comparisons if X is in any of 2^{k-1} positions in the list and if X is in any of the $n + 1$ gaps, so $s_k = 2^{k-1} + n + 1$. (If we did not assume $n = 2^k - 1$, only $k - 1$ comparisons might be done for some of the gaps; see Table 1.1 for $n = 25$.) The average number of comparisons done is

$$A(n) = \frac{1}{2n + 1} \sum_{t=1}^{k} t s_t = \frac{1}{2n + 1} \left[\sum_{t=1}^{k} t 2^{t-1} + k(n + 1) \right].$$

We can evaluate $\sum\limits_{t=1}^{k} t2^{t-1}$ as follows:

$$\sum_{t=1}^{k} t2^{t-1} = \sum_{t=1}^{k} t(2^t - 2^{t-1})$$

$$= \sum_{t=1}^{k} t2^t - \sum_{t=0}^{k-1} (t+1)2^t$$

$$= \sum_{t=1}^{k} t2^t - \sum_{t=0}^{k-1} t2^t - \sum_{t=0}^{k-1} 2^t$$

$$= k2^k - (2^k - 1) = (k-1)2^k + 1.$$

So

$$A(n) = \frac{(k-1)2^k + 1}{2n+1} + \frac{k(n+1)}{2n+1}$$

$$= \frac{(k-1)2^k + 1}{2 \cdot 2^k - 1} + \frac{k2^k}{2 \cdot 2^k + 1}$$

$$\approx \frac{k-1}{2} + \frac{k}{2} = k - \frac{1}{2} = \lfloor \log n \rfloor + \frac{1}{2}.$$

Thus we have proved the following theorem.

Theorem 1.3 BINARY SEARCH (Algorithm 1.4) does approximately $\lfloor \log n \rfloor + \frac{1}{2}$ comparisons on the average for lists with n entries.

OPTIMALITY

We will show that the binary search algorithm is optimal in the class of algorithms that can do no other operations on the list entries except comparisons. We will establish a lower bound on the number of comparisons needed by examining *decision trees* for algorithms in the class. Let A be such an algorithm. A decision tree for A and a given input size n is a binary tree whose nodes are labeled with numbers between 1 and n and are arranged according to the following rules:

1. The root of the tree is labeled with the index of the first entry in the list to which the algorithm A compares X.

2. Suppose the label on a particular node is i. Then the label on the left child of that node is the index of the entry to which the algorithm will compare X next if $X < L(i)$. The label on the right child is the index of the entry to which the algorithm will compare X next if $X > L(i)$. The node does not have a left (or right) child if the algorithm halts after comparing X to $L(i)$ and discovering that $X < L(i)$ [or $X > L(i)$]. (There is no branch for the case $X = L(i)$. A reasonable algorithm would simply halt in that case.)

The algorithm A may do a lot of other work that is not reflected in the decision tree; it may, for example, compare two elements in the list even though it is

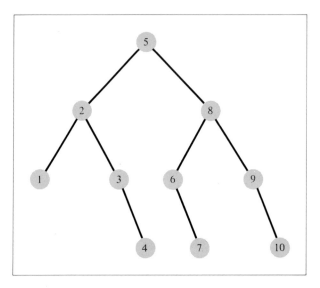

Figure 1.1
Decision tree for the binary search algorithm with $n = 10$.

pointless to do so. The class of algorithms considered is very broad and includes inefficient as well as efficient ones. Figure 1.1 shows the decision tree for the binary search algorithm with $n = 10$.

Given a particular input, algorithm A will perform the comparisons indicated along one particular path from the root of its decision tree. The number of basic operations performed is the number of nodes on the path. The number of comparisons performed in the worst case is the number of nodes on the longest path from the root to a leaf; i.e., 1 + the depth of the tree. Thus to establish a lower bound of, say, $G(n)$, for some function G, on the number of comparisons needed in the worst case, we prove that the decision tree for A has depth at least $G(n) - 1$.

Let d be the depth of the decision tree for A. Lemma 0.2 says that there are at most $2^{d+1} - 1$ nodes in the tree. We want to relate the depth to n, the input size. We claim that there are at least n nodes in the tree. Suppose, to the contrary, that there is no node labeled i, for some i between 1 and n. We can make up two input lists L and L' such that for $1 \leqslant j \leqslant n$ and $j \neq i$, $L(j) = L'(j) \neq X$ and $L(i) = X$ but $L'(i) \neq X$. Since no node in the decision tree is labeled i, the algorithm A never compares X to $L(i)$ or $L'(i)$; thus it behaves the same way on both inputs since their other entries are identical, and it must give the same output for both. Thus A gives the wrong output for at least one of the lists and it is not a correct algorithm. We can conclude that $n \leqslant$ the number of nodes in the tree. So $n \leqslant 2^{d+1} - 1$ and $\log(n + 1) \leqslant d + 1$. Since d is an integer, $d \geqslant \lceil \log(n + 1) \rceil - 1$ and the number of comparisons done by A in the worst case is at least $\lceil \log(n + 1) \rceil = \lfloor \log n \rfloor + 1$. Since A was an arbitrary algorithm from the class of algorithms considered, we have proved the following theorem.

Theorem 1.4 Any algorithm to find X in a list of n entries (by comparing X to list entries) must do at least $\lfloor \log n \rfloor + 1$ comparisons for some input.

Since Algorithm 1.4 does $\lfloor \log n \rfloor + 1$ comparisons in the worst case, it is optimal.

1.5   *Finding the Largest and Second Largest Entries in a List*

In Example 1.3 we considered the problem of finding the largest entry in a list with n entries and concluded that $n - 1$ comparisons of list entries are both necessary and sufficient to solve the problem. In this section we will study the problem of finding the second largest entry in a list. Throughout this section we will use the names M and S to refer to the maximum and second largest entries, respectively. We will use the terminology of contests, or tournaments, in discussing the results of comparisons. The comparand that is found to be larger will be called the winner; the other will be called the loser. For simplicity in describing the problem and algorithms, we will assume that the list entries are distinct.

S can be found by first finding M using Algorithm 1.3 and then eliminating M from the list and using Algorithm 1.3 again to find the largest entry in the resulting list of $n - 1$ entries. Thus S can be found by doing $(n - 1) + (n - 2)$, or $2n - 3$, comparisons. Is there a better way? Observe that some of the information obtained by the algorithm while finding M can be used to decrease the number of comparisons performed in finding S. Specifically, any entry that loses to an entry that is not the maximum cannot possibly be the second largest. All such entries discovered while finding M may be ignored during the second pass through the list. (Keeping track of them is a problem that will be considered later.)

Using the algorithm in Example 1.3 on a list with five entries, the results might be as follows:

Comparands	Winner
$L(1), L(2)$	$L(1)$
$L(1), L(3)$	$L(1)$
$L(1), L(4)$	$L(4)$
$L(4), L(5)$	$L(4)$

Then $M = L(4)$ and S is either $L(5)$ or $L(1)$ because both $L(2)$ and $L(3)$ lost to $L(1)$. Different sets of specific numbers can be chosen to make either $L(5)$ or $L(1)$ the second largest entry. In this case, finding S requires one more comparison.

It may happen, however, that during the first pass through the list to find M we do not obtain any information useful for finding S. The worst case occurs when M is $L(1)$; each other entry would be compared only to M. Does this mean that in the worst case $2n - 3$ comparisons must be done to find S? Not necessarily. In the discussion above we have been using a specific algorithm to find the maximum entry. We know from Section 1.3 that no algorithm can find M by doing less than $n - 1$ comparisons, but another algorithm for finding M may provide more information useful for eliminating some entries during the second pass through the list. The tournament method described next provides such information.

THE TOURNAMENT METHOD

The tournament method derives its name from the fact that it performs comparisons in the same way that tournaments are played. Elements are paired off and compared in what may be called the first round. In succeeding rounds, the winners from the preceding round are paired off and compared. (If at any round there is an odd number of entries, one can be arbitrarily designated as a winner and not compared to another in that round.) Ultimately every entry except one will have lost to some other entry. A tournament can be described by a tree diagram as shown in Fig. 1.2. Each list entry is a leaf, and at each level the nodes are paired and compared and the parent of each pair is the winner. The root will be the largest entry.

In the process of finding M, every entry except M loses in one comparison. We have observed that only entries that lose to M may be S. Thus if we can compute the number of entries that lose to M we will know how many comparisons must be done

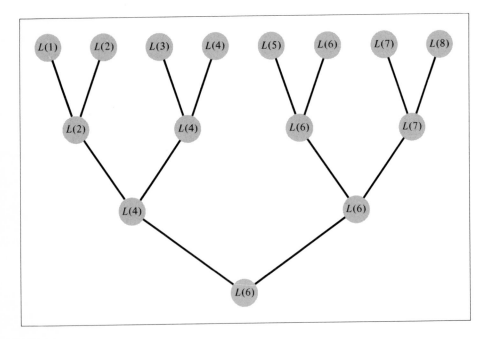

Figure 1.2
An example of a tournament $M = L(6)$; S may be $L(4)$, $L(5)$, or $L(7)$.

on the second pass through the list. Roughly half the entries in one round will be losers and will not appear in the next. If n is a power of 2 there are exactly $\log n$ rounds; in general, the number of rounds is $\lceil \log n \rceil$. M is involved in at most one comparison in each round so there are at most $\lceil \log n \rceil$ entries that may be S. The method of Algorithm 1.3 can be used to find the largest of these $\lceil \log n \rceil$ entries by doing $\lceil \log n \rceil - 1$ comparisons. Thus the tournament finds M and S by doing a total of $n + \lceil \log n \rceil - 2$ comparisons. This is an improvement over our first result of $2n - 3$. Can we do better?

LOWER BOUND FOR FINDING S

The problem stated at the beginning of this section was to find the second largest entry in a list. Our scheme in both of the methods considered was to find the maximum, doing $n - 1$ comparisons, and then to find the maximum of some subset of the remaining entries. We are not explicitly told to find M, but we do. This is not wasted effort. Any algorithm that finds S must also find M because to know that an entry is the second largest, one must know that it is not the largest; that is, it must have lost in one comparison. The winner of the comparison in which S loses must, of course, be M. This argument gives a lower bound on the number of comparisons needed to find S, namely $n - 1$, because we already know that at least $n - 1$ comparisons must be done by any algorithm that finds M. But one would expect that this lower bound could be improved because an algorithm to find S should have to do more work than an algorithm to find M. We will prove the following theorem, which has as a corollary that the tournament method is optimal. The class of algorithms considered consists of those that can do no operations on list entries other than comparing or copying them.

Theorem 1.5 Any algorithm to find the second largest entry in a list of n entries must do at least $n + \lceil \log n \rceil - 2$ comparisons in the worst case.

Proof. Finding M requires that each entry other than M be a loser in a comparison; thus (since finding S requires finding M), there are $n - 1$ comparisons with distinct losers. If M was a comparand in $\lceil \log n \rceil$ of these comparisons, then all but one of the $\lceil \log n \rceil$ entries that lost to M must lose again for S to be correctly determined. Then a total of at least $n + \lceil \log n \rceil - 2$ comparisons would be done. Therefore we will show that for any algorithm to find S there is an input list for which M is compared to $\lceil \log n \rceil$ distinct entries.

    Given a particular algorithm to find S, we construct an input list consistent with the rules given below while carrying out the instructions in the algorithm. An entry is said to be undefeated at any point if it has not yet lost in any comparison. (Entries that have not yet been in any comparison are undefeated.) An entry is called a loser if it has lost at least one comparison. For each entry we keep a count of wins against previously undefeated entries. The rules for determining results of comparisons follow.

Comparands	*Results*
1. One undefeated, one loser	The undefeated wins.
2. Both undefeated	The winner is the entry with the most previous wins. If they have the same number of wins, select either as the winner.
3. Both losers	The winner is determined by values already assigned to the entries.

It is always possible to choose entries for a list consistent with these rules because the only winners explicitly specified are undefeated entries; it is always possible to make the value of such an entry arbitrarily large (to force it to win the current comparison) without contradicting the results of any of the previous comparisons.

Example To illustrate the use of these rules, we construct an input list with five entries for an algorithm A that finds S. (The algorithm A will not be completely described. At each step we specify only the comparison it does next based on the results of the previous ones.) Elements in the list that have not yet been chosen are denoted by asterisks. Thus initially the entries are *, *, *, *, *. Suppose A first compares $L(1)$ and $L(2)$. Both are undefeated with the same number of wins (zero), so, by rule 2, arbitrary values can be inserted. We let $L(1) = 20$ and $L(2) = 10$. $L(1)$ now has one win and the list is 20, 10, *, *, *. Next A compares $L(1)$ and $L(3)$. Both are undefeated but $L(1)$ has more wins so it wins again. Let $L(3)$ be 15. The list is now 20, 10, 15, *, *, and $L(1)$ has two wins. Next A compares $L(4)$ and $L(5)$. By rule 2, we can assign arbitrary values. Let $L(4) = 30$ and $L(5) = 40$. Thus $L(5)$ has one win and the list is 20, 10, 15, 30, 40. Next A compares $L(1)$ and $L(5)$. Both are undefeated but $L(1)$ has more wins so by rule 2 it is to be the winner. With the current values $L(1) < L(5)$, but this can be corrected by increasing $L(1)$ to, say, 50. The list is 50, 10, 15, 30, 40. At this point the algorithm has found the largest entry, $L(1)$, and continues to find S. Suppose it compares $L(2)$ and $L(3)$. By rule 3, the previously assigned values determine the result: $L(3) > L(2)$. Finally A compares $L(3)$ and $L(5)$. Again both are losers and $L(5) > L(3)$. The algorithm terminates with $M = L(1)$ and $S = L(5)$. It has performed $6 = 5 + \lceil \log 5 \rceil - 2$ comparisons.

We now show that if the data are consistent with the rules given above, the algorithm will compare M to at least $\lceil \log n \rceil$ distinct entries. We construct a sequence of forests whose nodes are the list entries. To start, each entry is the root of a separate tree. Undefeated entries will always be roots, and vice versa. Whenever the algorithm compares two entries that were previously undefeated, the loser's tree is attached as a subtree of the winner. There is no change in the trees when either comparand is a loser. Figure 1.3 shows the forests constructed for the example above.

Lemma 1.6 If at some point during the construction of the trees X is undefeated and has been in q comparisons against distinct previously undefeated entries, then X is the root of a tree with at most 2^q nodes.

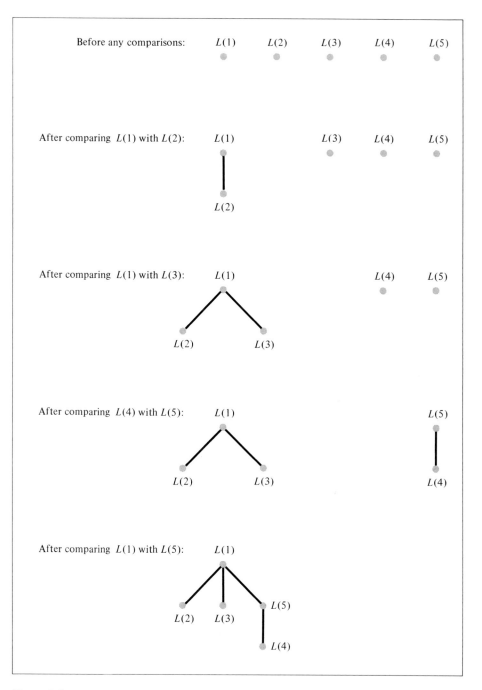

Figure 1.3
Forests for the input constructed in the proof of Theorem 1.5.

Proof, by induction on q. If $q = 0$, X has not been compared to any other undefeated entry and X itself is undefeated so it is the root of the tree consisting of itself alone. This tree has one, or 2^0, node.

Suppose that for any p such that $0 \leqslant p < q$, any undefeated entry that has been compared to p previously undefeated entries is the root of a tree with at most 2^p nodes. Suppose X is undefeated and was in q such comparisons, and that X was compared to Y in the last such comparison. Then, before the comparison with Y, X was the root of a tree with at most 2^{q-1} nodes. Since Y lost to X, by rule 2 Y had at most $q - 1$ previous wins, so Y was the root of a tree with at most 2^{q-1} nodes. After X and Y were compared, the tree with root Y was attached to the tree with root X, making X the root of a tree with at most $2^{q-1} + 2^{q-1} = 2^q$ nodes. This proves the lemma.

To complete the proof of Theorem 1.5 observe that since any node that is a root is undefeated, when the algorithm terminates M must be the root of a tree with all n entries as nodes. Let q be the total number of comparisons in which M and a previously undefeated entry were compared. Then, by the lemma, $n \leqslant 2^q$, so $\log n \leqslant q$, and since q is an integer, $q \geqslant \lceil \log n \rceil$. □

IMPLEMENTATION OF THE TOURNAMENT METHOD FOR FINDING M AND S

To conduct the tournament to find M we need a means of keeping track of the winners in each round. This can be done with an extra array of pointers or by careful indexing if the entries may be moved so that the winner is always placed in the higher indexed cell of the two being compared. We leave the choice and the details to the reader.

After M has been found by the tournament, only those entries that lost to it are to be compared to find S. How can we keep track of the elements that lose to M when we do not know in advance which entry is M? One way is to maintain linked lists of entries that lost to each undefeated entry. This can be done by allocating an array for links indexed to correspond to the list entries. Initially all links would be null. After each comparison in the tournament, the entry that lost would be added to the winner's linked list. Specifically, suppose that $L(i)$ and $L(j)$ are compared and $L(i)$ wins. Then the instructions for updating $L(i)$'s loser list would be:

$$\text{LINK}(j) \leftarrow \text{LINK}(i)$$

$$\text{LINK}(i) \leftarrow j$$

Figure 1.4 shows the LINK array after each of the first five comparisons of a tournament on six items. All links were initially Λ. Old values of LINK(1), LINK(3), and LINK(5) are shown as crossed-out entries. (If entries in the list are rearranged by the tournament algorithm as suggested in the previous paragraph, then the links must be set differently. If $L(i)$ wins a comparison against $L(j)$ and the tournament interchanges the two, then LINK(j) should be set to i, and LINK(i) is unchanged.)

When the tournament is complete and M has been found, it is easy to find S by traversing the linked list of losers to M.

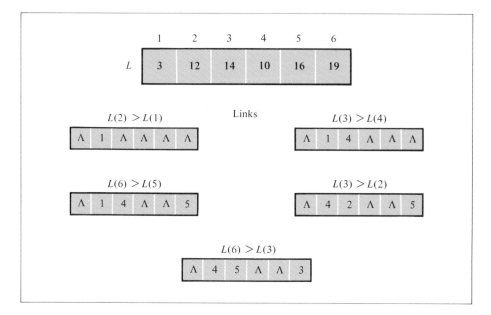

Figure 1.4
Link values for a tournament on six items. The loser list for $L(6)$ ultimately contains $L(3)$ and $L(5)$.

TIME AND SPACE REQUIREMENTS

The tournament method for finding M and S uses extra space for links; the number of extra cells needed is n, or at most $2n$ if pointers are used to control the tournament. The running time of the algorithm is $\Theta(n + \lceil \log n \rceil - 2) = \Theta(n)$ since the number of operations done to set and use the links is roughly proportional to the number of comparisons done.

We can find the largest and second largest entries in a list by using FINDMAX twice doing $2n - 3$ comparisons, or we can use the more complicated tournament method doing $n + \lceil \log n \rceil - 2$ comparisons at most. Which method is better? The results of Exercise 1.16 should be instructive. Observe that both functions are $\Theta(n)$. If comparisons take more time on a particular computer than most of the book-keeping instructions, the tournament method will run faster, but generally this is not the case. Since the tournament method does more instructions per comparison while finding M, it may well be slower.

1.6   *Exercises*

Section 1.3: Analyzing Algorithms

1.1. Let α and β be real numbers such that $0 < \alpha < \beta$. Verify that n^{α} is $0(n^{\beta})$ but n^{β} is not $0(n^{\alpha})$.

1.2. a) Group the following functions so that f and g are in the same group if and only if $f = \Theta(g)$.

   b) List the groups formed in (a) from lowest order to highest.

n	n^3
2^n	$\log n$
$n \log n$	$n^{1/2} + \log n$
$n - n^3 + 7n^5$	$(\log n)^2$
$n^2 + \log n$	$n!$
n^2	$\ln n$

1.3. a) Write an algorithm to find the median of three integers a, b, and c.

   b) Describe D, the set of inputs for your algorithm, in light of the discussion following Example 1.1.

   c) How many comparisons does your algorithm do in the worst case? On the average?

   d) How many comparisons are necessary in the worst case to find the median of three numbers? Justify your answer.

1.4. Suppose the following algorithm is used to evaluate the polynomial $p(X) = a_n X^n + a_{n-1} X^{n-1} + \cdots + a_1 X + a_0$.

```
p ← a₀
XPOWER ← 1
for i ← 1 to n do
      XPOWER ← X*XPOWER
      p ← p + aᵢ*XPOWER
end
```

   a) How many multiplications are done in the worst case? How many additions?

   b) How many multiplications are done on the average?

Section 1.4: Searching an Ordered List

1.5. Write an algorithm to find X in an ordered list by the method suggested on pp. 32–33, that is, by comparing X to every fourth entry until X itself or an entry larger than X is found, and then, in the latter case, searching for X among the preceding three. How many comparisons does your algorithm do in the worst case?

1.6. Describe the decision tree for the SEQUENTIAL SEARCH algorithm (Algorithm 1.1) in Section 1.3 for an arbitrary n.

1.7. Draw a decision tree for the algorithm in Exercise 1.5 with $n = 17$.

1.8. Verify that $\lceil \log (n + 1) \rceil = \lfloor \log n \rfloor + 1$.

1.9. Find a formula for $A(n)$, on p. 32, without summations.

1.10. How can you modify BINARY SEARCH (Algorithm 1.4) to eliminate unnecessary work if you are certain that X is in the list? Draw a decision tree for the modified algorithm for $n = 7$. Do a worst-case and average-behavior analysis. (For the average, you may assume $n = 2^k - 1$ for some k.)

1.11. Let S be a set of m integers and let L be a list of n integers ($n \leqslant m$) randomly chosen from the set S. Assume that the entries in L are sorted in ascending order. Let x be an element of S. On the average, how many comparisons will be done by BINARY SEARCH (Algorithm 1.4) given L and x as input? Express your answer as a function of n and m.

1.12. Suppose that the function Q is defined for all powers of 2 and is described by the following recurrence relation and boundary condition:

$$Q(n) = n - 1 + 2Q\left(\frac{n}{2}\right)$$

$$Q(1) = 0$$

Find a closed form for Q.

Section 1.5: Finding the Second Largest Entry in a List

1.13. Write an algorithm in detail for the tournament method to find M and S.

1.14. How many comparisons are done by the tournament method to find S on the average if n is a power of 2?

1.15. For certain values of n and certain inputs, the tournament method can find both M and S after doing only $n - 1$ comparisons. Give an example to illustrate this statement.

1.16. Write assembly language routines (without I/O) to find M and S by the tournament method and by using Algorithm 1.3 twice. Using the manuals for your computer to find the execution times for each instruction, estimate the worst-case execution time for each program.

PROGRAMS

1. Exercise 1.13.

NOTES AND REFERENCES

Michael O. Rabin's Turing Award Lecture (Rabin, 1977) gives the reader a very nice overview of questions, techniques, and points of view studied in computational complexity. Weide (1977) is a particularly good survey of analysis techniques and contains a useful glossary and an extensive bibliography. The reader may find it useful to skim now and read more thoroughly after completing several chapters of this book.

Several other texts on design and analysis of algorithms are listed in the references.

Knuth (1976) discusses the meaning and history of the notations $O(f)$ and $\Theta(f)$.

Elpas, Levitt, Waldinger, and Waksman (1972) and Hantler and King (1976) are surveys of both formal and informal techniques for proving program correctness. Dahl,

Dijkstra, and Hoare (1972), particularly the first chapter, is a good discussion of program design and correctness.

The problems studied in Sections 1.4 and 1.5 are also covered in Knuth (1973), which provides more references and some historical background. Section 1.5 considers a special case of the problem of finding the tth largest entry in a list. For the general problem, see Knuth (1973) or Aho, Hopcroft, and Ullman (1974).

The reader who wishes to browse through research articles will find a lot of material in the *Journal of the ACM,* the *Proceedings of the ACM Symposium on Theory of Computing* (annual), *SIGACT News,* the *SIAM Journal on Computing, Transactions on Mathematical Software,* and the *Communications of the ACM,* to name a few sources.

Knuth (1977), a paper about the space complexity of songs, is very highly recommended for comic relief.

The number of possible chess games is discussed in Good (1968), particularly in Appendixes D and E.

Sorting 2

$$2$$

2.1 *Introduction*

In this chapter we will study several algorithms for sorting, that is, for rearranging the elements of a file or list into order. We assume that each item in the file to be sorted contains an identifier, called a *key*, which is an element of some linearly ordered set, and that two keys can be compared to determine which is larger or that they are equal. We will always sort keys into nondecreasing order. Each entry in the file may contain other information aside from the key. For example, the keys in a telephone book are the names of people or businesses who have telephones; the information associated with a key is the address and phone number of that person or business. In a dictionary entry the key is the word to be defined; the associated information is the definition, pronunciation guide, and whatever else the dictionary tells about the word. When keys are rearranged during the sorting process the associated information will also be rearranged as appropriate, but most of the time we will refer only to the keys and make no explicit mention of the rest of the entry.

There are several good reasons for studying sorting algorithms. First, they are of practical use because sorting is done extremely often; a very large percentage of the running time of computers is spent on sorting. Just as having the entries in telephone books and dictionaries in alphabetical order makes them easy to use, working with large sets of data in computers is facilitated when the data are sorted. Second, quite a lot of sorting algorithms have been devised (more than will be covered here), and studying a number of them should impress upon the reader the fact that one can take many different points of view toward the same problem. The discussion of the algorithms in this chapter should provide some insights on the questions of how to improve a given algorithm and how to choose among several. Third, sorting is one of few problems for which we can easily derive good lower bounds for the worst case and average behavior of the algorithms. The bounds are good in the sense that there are algorithms that do approximately the minimum amount of work specified.

The algorithms considered in Sections 2.2 through 2.6 are all from the class of sorting algorithms that may compare keys (and copy them) but cannot do other operations on the keys. We will call these "algorithms that sort by comparison of keys." The measure of work used for analyzing algorithms in this class is the number of comparisons of keys. In Section 2.4 lower bounds on the number of comparisons performed by such algorithms are established. Section 2.7 discusses sorting algorithms for which different measures of work are appropriate. The algorithms in Sections 2.2 through 2.7 are called *internal sorts* because the data are assumed to be in the computer's high-speed, random-access memory. In Section 2.9 we study an algorithm for sorting large sets of data stored on external, slower storage devices with restrictions on the way in which data are accessed. Such algorithms are called *external sorts.* Section 2.8 discusses the problem of merging sorted files. Merging has many applications, one of which is the external sort discussed in Section 2.9.

The reader should work on Exercise 2.1 before proceeding.

2.2   *BUBBLESORT*

BUBBLESORT is one of several straightforward sorting algorithms. It sorts by making several passes through the file comparing pairs of keys in adjacent locations and interchanging them if they are out of order. That is, the first and second keys are compared and interchanged if the first is larger than the second; then the (new) second and the third keys are compared and interchanged if necessary, and so on. It is easy to see that the largest key will filter down to the bottom; on subsequent passes it will be ignored. If on any pass no entries are interchanged, the file is completely sorted and the algorithm can halt. For this reason, on each pass through the file a flag is set to indicate whether or not any interchanges were done. Algorithm 2.1 makes this informal description of the method precise. It is not our final version of BUBBLESORT, though; it will be improved below.

Algorithm 2.1 BUBBLESORT (to be improved below)

Input: L, an array of keys, and $n \geq 0$, the number of keys.

Output: L with keys in nondecreasing order.

1. $k \leftarrow n$; FLAG $\leftarrow 1$
      [k is the number of pairs to be compared, and FLAG > 0 indicates that there is more work to be done.]
2.   **while** FLAG > 0 **do**
3. $k \leftarrow k - 1$

```
4.          FLAG ← 0
5.          for j ← 1 to k do
6.              if L(j) > L(j + 1) then do
7.                              L(j) ↔ L(j + 1)
8.                              FLAG ← 1
                            end
        end
    end
```

The example in Fig. 2.1 illustrates how BUBBLESORT works.

Algorithm 2.1 avoids some unnecessary comparisons by terminating as soon as it discovers that the file is sorted. We can improve the algorithm to avoid unnecessary comparisons in other situations as well. Suppose, for example, $n = 20$ and the input file consists of the keys in Fig. 2.1, that is, 8, 3, 4, 9, 7, followed by fifteen numbers all larger than 9 and already sorted. The algorithm will terminate after three passes through the file because three passes are needed to put the first five entries in order; however, a total of 54 comparisons will be done and most of the already sorted keys will be examined three times when one pass would be sufficient to discover that they are already sorted. We use the fact that if on any pass the last interchange occurs at the jth and $(j + 1)$st positions, then each entry in the tail of the file – that is, positions $j + 1$ through n – is in its proper place. The unnecessary work is avoided by setting k to the last value of j for which an interchange may be necessary. FLAG can be used to keep track of this index. Thus our BUBBLESORT algorithm is as follows:

Algorithm 2.2 BUBBLESORT
Algorithm 2.1 with the following substitutions:

Line 1:	FLAG ← n
Line 3:	k ← FLAG − 1
Line 8:	FLAG ← j

The test in line 2 was phrased to work correctly with or without the changes.

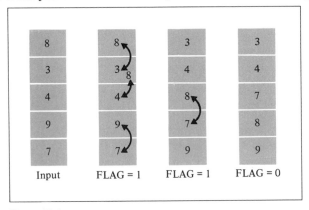

Figure 2.1
BUBBLESORT (using Algorithm 2.1).

WORST-CASE ANALYSIS

The modification just described does not affect the worst-case behavior of the algorithm; it was designed to eliminate work for some partially sorted inputs. So our worst-case analysis is for both Algorithms 2.1 and 2.2.

The instructions in the **for** loop may be executed for each value of k from $n-1$ down to 1. The comparison in line 6 is executed k times for each value of k. Thus in the worst case the number of comparisons is

$$\sum_{k=1}^{n-1} k = \frac{n(n-1)}{2} \approx \frac{n^2}{2}$$

and BUBBLESORT is an $\Theta(n^2)$ sort.

Aside from comparisons, another important operation performed by BUBBLE-SORT is the interchanging of two keys in line 7. A key may be moved from one location to another many times before it finally reaches its correct place. Thus to get a more precise idea of how much work BUBBLESORT does, we can count the number of exchanges as well as the number of comparisons. In the worst case, every comparison results in an exchange yielding a total of $n(n-1)/2$.

AVERAGE BEHAVIOR

In many applications the keys to be sorted are distinct. For the average-behavior analysis we will assume that this is the case; in other words, we assume that there is zero probability that an input file will contain duplicate keys.

A *permutation* on n letters is a one-to-one function from the set $N = \{1, 2, \ldots, n\}$ onto itself. There are $n!$ distinct permutations on n letters. Let the keys in the unsorted file L be $x_1, x_2, \ldots, x_n$. There is a permutation α such that for $1 \le i \le n$, $\alpha(i)$ is the correct position of x_i when the file is sorted. Without loss of generality, we can assume that the keys are the integers $1, 2, \ldots, n$ since we can substitute 1 for the smallest key, 2 for the next smallest, and so on, without causing any changes in the instructions carried out by the algorithm. Then the unsorted input file is $\alpha(1)$, $\alpha(2), \ldots, \alpha(n)$. For example, consider the input file 2, 4, 1, 5, 3. $\alpha(1) = 2$ means that the first key, 2, belongs in the second position, which it clearly does. $\alpha(2) = 4$ because the second key, 4, belongs in the fourth position, and so on. We will identify the permutation α with the file $\alpha(1), \alpha(2), \ldots, \alpha(n)$. To compute the average number of comparisons done by a sorting algorithm we try to find out how many comparisons the algorithm does for each permutation.

An *inversion* of the permutation α is a pair $(\alpha(i), \alpha(j))$ such that $i < j$ and $\alpha(i) > \alpha(j)$. If $(\alpha(i), \alpha(j))$ is an inversion, the ith and jth keys in the file are out of order relative to each other. For example, the permutation 2, 4, 1, 5, 3 has four inversions $(2, 1)$, $(4, 1)$, $(4, 3)$, and $(5, 3)$. Each time a comparison in the BUBBLE-SORT algorithm results in an exchange of two keys, exactly one inversion is removed from the list. Since some comparisons do not result in an exchange, the number of comparisons performed by BUBBLESORT on the input $\alpha(1), \alpha(2), \ldots, \alpha(n)$ is at least the number of inversions of α. Therefore, the average number of inversions in permutations on n letters is a lower bound for the average number of comparisons done by BUBBLESORT, so we investigate inversions.

For any permutation α, the sequence $B(\alpha) = (b_1, b_2, \ldots, b_n)$ is defined by letting b_i be the number of elements to the right of i that are less than i. For the permutation 2, 4, 1, 5, 3, $b_1 = b_3 = 0$, $b_2 = b_5 = 1$, and $b_4 = 2$. Observe that b_i is simply the number of inversions of α whose first component is i, and the total number of inversions is $\Sigma_{i=1}^n b_i$. Also, for $1 \leqslant i \leqslant n$, $0 \leqslant b_i \leqslant i - 1$. If we are given any sequence $B = (b_1, b_2, \ldots, b_n)$ such that $0 \leqslant b_i \leqslant i - 1$ for $1 \leqslant i \leqslant n$, it is possible to find a unique permutation α such that $B = B(\alpha)$ (see Exercise 2.6). Thus since there is a one-to-one correspondence between such sequences and permutations, we conclude that for each nonnegative integer k, the number of permutations with exactly k inversions is the number of sequences $b_1, b_2, \ldots, b_n$ with $0 \leqslant b_i \leqslant i - 1$ such that $\Sigma_{i=1}^n b_i = k$. We denote this number by $s(k)$. A permutation on n letters can have at most $[n(n-1)]/2$ inversions, so the average number of inversions is

$$\frac{1}{n!} \sum_{k=0}^{n(n-1)/2} k \cdot s(k).$$

Using techniques of combinatorics, the value of $s(k)$ and then the value of the summation can be computed. The result is $[n(n-1)]/4$, or approximately $n^2/4$. (See the notes and references at the end of the chapter.)

Let $A(n)$ be the average number of comparisons done by BUBBLESORT on files of size n, assuming all permutations of the data are equally likely. Then $A(n) \geqslant [n(n-1)]/4 = \Theta(n^2)$. Thus both the worst-case and average behavior of BUBBLE-SORT is $\Theta(n^2)$. The discussion in the previous paragraphs tell us that *any* sorting algorithm that interchanges only adjacent keys, removing at most one inversion after each comparison, will do at least $\Theta(n^2/4)$ comparisons on the average and therefore at least $\Theta(n^2)$ in the worst case. If we want significantly faster algorithms we will have to move keys more than one position at a time.

2.3   QUICKSORT

THE ALGORITHM

QUICKSORT is a recursive sorting algorithm. Roughly speaking, it rearranges the keys and splits the file into two subsections, or subfiles, such that all keys in the first section are smaller than all keys in the second section. Then QUICKSORT sorts the two subfiles recursively (i.e., by the same method), with the result that the entire file is sorted.

Let L be the array of keys and let k and m be the indexes of the first and last entries, respectively, in the subfile QUICKSORT is currently sorting. (Initially $k = 1$

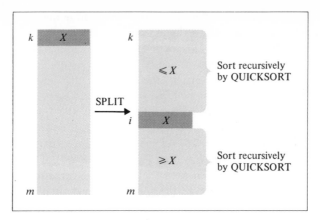

Figure 2.2
QUICKSORT.

and $m = n$.). The SPLIT algorithm chooses a key X from the subfile and rearranges the entries, finding an integer i such that for $k \leqslant r < i$, $L(r) \leqslant X$; $L(i) = X$; and for $i < r \leqslant m$, $L(r) \geqslant X$. X is then in its correct position and is ignored in the subsequent sorting. (See Fig. 2.2.)

QUICKSORT is described by the following recursive algorithm.

Algorithm 2.3 QUICKSORT (L, k, m)

if $k < m$ **then do**   SPLIT(L, k, m, i)
                       QUICKSORT$(L, k, i - 1)$
                       QUICKSORT$(L, i + 1, m)$
       end

The SPLIT routine may choose as X any key in the file between $L(k)$ and $L(m)$; for simplicity, let $X = L(k)$. An efficient splitting algorithm uses two pointers, i and j, initialized to k and $m + 1$, respectively, and begins by copying X elsewhere so that the position $L(i)$ is available for some other entry. The "hole," or empty location, at $L(i)$ is filled by decrementing j until $L(j) < X$, and then copying $L(j)$ into $L(i)$. Now the hole is at $L(j)$ and can be filled by incrementing i until $L(i) > X$, and then copying $L(i)$ into $L(j)$. This procedure continues until the values of i and j meet; then X is inserted in the final hole. We assume SPLIT works this way and returns the final value of i. Writing out the algorithm in detail is left as an exercise. Observe that SPLIT compares each key except the key originally in $L(k)$ to X, so it does $m - k$ comparisons. Any other reasonable partitioning algorithm will do the same number of comparisons. The advantage of SPLIT is that every entry that must be moved is moved only once (except X, which is moved twice).

Observe that the QUICKSORT routine provides the control structure for the sort, but all the work of comparing and rearranging entries is done by the nonrecursive routine SPLIT. In practice, it is better not to write QUICKSORT as a recursive program; the programmer can employ some time- and space-saving tricks by manipulating the stack in the program rather than leaving it to the compiler for a recursive programming language. Some of these tricks and other improvements will be discussed after we analyze the algorithm.

WORST CASE

If when SPLIT is executed $L(k)$ is the largest key in the current subfile (that is, $L(k) \geqslant L(r)$ for $k \leqslant r \leqslant m$), then SPLIT will move it to the bottom [to position $L(m)$] and partition the file into one section with $m - k$ entries (all but the bottom one) and one section with no entries. All that has been accomplished is moving the maximum entry to the bottom. Similarly, if the smallest entry in the file is in position $L(k)$, SPLIT will simply separate it from the rest of the list, leaving $m - k$ items still to be sorted. Thus if the input is arranged so that each time SPLIT is executed, $L(k)$ is the largest (or the smallest) entry in the section being sorted, then, if we let $s = m - k + 1$, the number of keys in the unsorted section, the number of comparisons done is

$$\sum_{s=2}^{n} (s-1) = \frac{n(n-1)}{2}.$$

An input file can be arranged to force this number of comparisons (Exercise 2.10) and the algorithm never need do more, so in the worst case QUICKSORT does $[n(n-1)]/2$ comparisons, the same as BUBBLESORT. Is the name QUICKSORT just a bit of false advertising?

AVERAGE BEHAVIOR

When studying BUBBLESORT we observed that if a sorting algorithm removes at most one inversion from the permutation of the keys after each comparison, then it must do at least $(n^2 - n)/4$ comparisons on the average. QUICKSORT, however, does not have this restriction. The SPLIT algorithm can move keys across a large section of the entire file, eliminating up to $n - 2$ inversions at one time. QUICKSORT deserves its name because of its average behavior.

Consider a situation in which QUICKSORT works quite well. Suppose that each time SPLIT is executed, it partitions the file into two roughly equal subfiles. To simplify the computation, assume that $n = 2^p - 1$ for some p. The number of comparisons done by QUICKSORT on a file with n entries under these assumptions is described by the recurrence relation

$$Q(p) = 2^p - 2 + 2Q(p-1)$$

$$Q(1) = 0.$$

The first term in $Q(p)$, $2^p - 2$, is $n - 1$, the number of comparisons done by SPLIT the first time. The second term is the number of comparisons done by QUICKSORT to sort the two subfiles, each of which has $(n-1)/2$, or $2^{p-1}-1$, entries. Expand the recurrence relation to get

$$Q(p) = 2^p - 2 + 2Q(p-1) = 2^p - 2 + 2(2^{p-1}-2) + 4Q(p-2)$$
$$= 2^p - 2 + 2^p - 4 + 2^p - 8 + 8Q(p-3).$$

Thus

$$Q(p) = \sum_{i=1}^{p-1} (2^p - 2^i) = (p-1)2^p - \sum_{i=1}^{p-1} 2^i$$
$$= (p-1)2^p - (2^p - 2) = \lfloor \log n \rfloor (n+1) - n + 1.$$

Thus if $L(k)$ were close to the median each time the file is split, the number of comparisons done by QUICKSORT would be $\Theta(n \log n)$. This is significantly smaller than $\Theta(n^2)$. But if we assume all permutations of the input data are equally likely, are there enough cases for which QUICKSORT behaves well to affect the average? The analysis is complicated (see the notes and references at the end of the chapter), but the result is that, on the average, QUICKSORT does approximately $2n \log n$ comparisons.

SPACE USAGE

At first glance it may seem that QUICKSORT is an in-place sort. It is not. While the algorithm is working on one subfile, the beginning and ending indexes (call them the borders) of all the other subfiles yet to be sorted must be saved on a stack, and the size of the stack depends on the number of sublists into which the file will be split. This, of course, depends on n. In the worst case, SPLIT may split off one entry at a time in such a way that n pairs of borders are stored on the stack. Thus the amount of space used by the stack is proportional to n. One of the modifications to the algorithm described below can significantly reduce the maximum stack size.

IMPROVEMENTS ON THE BASIC QUICKSORT ALGORITHM

1.   We have seen that QUICKSORT works well if the key X used by SPLIT to partition a subfile is close to the median entry. Choosing $L(k)$ as X causes QUICKSORT to do poorly in cases where sorting should be easy. (See, for example, Exercise 2.11.) There are several other strategies for choosing X that decrease the chances of a worst case occurring. One is to choose a random integer q between k and m and let $X = L(q)$. Another is to let X be the median of the entries $L(k)$, $L[(k+m)/2]$, and $L(m)$. (In either case, the key in $L(k)$ would be moved to the position from which X was taken before proceeding with the SPLIT algorithm.) Both of these strategies require some extra work to choose X, but it pays off by improving the average running time of a QUICKSORT program.

2.   Every time SPLIT partitions a file, some data must be put on the stack. The manipulation of the stack takes time but is worthwhile when n is large because the algorithm is fast. For a very small file the time spent handling the stack and doing other bookkeeping outweighs the time gained by doing fewer comparisons than some

nonrecursive sort might require. QUICKSORT is not particularly good for small files. But, by the nature of the algorithm, for large n QUICKSORT will break the file up into small subfiles and recursively sort them. Thus whenever the size of a subfile is small, the algorithm becomes inefficient. This problem can be remedied by choosing a small M and sorting subfiles of size $\leq M$ by some other nonrecursive procedure, called OTHERSORT in the modified algorithm.

Algorithm 2.4 QUICKSORT (L, k, m)

if $m - k \geq M$ then do   SPLIT(L, k, m, i)
                            QUICKSORT$(L, k, i - 1)$
                            QUICKSORT$(L, i + 1, m)$
                  end
       else   OTHERSORT(L, k, m)

What value should M have? The best choice of M depends on the particular implementation of the algorithm (that is, the computer being used and the details of the program), since we are cutting bookkeeping time at the expense of more comparisons. A careful analysis of the program is needed to find the best value. In lieu of the analysis, a value close to 10 may do reasonably well.

3.   In a recursive implementation of QUICKSORT, the instructions QUICKSORT $(L, k, i - 1)$ and QUICKSORT$(L, i + 1, m)$ will each cause L, k, m, and i to be stacked. Much of the pushing and popping that will be done is unnecessary. After SPLIT, the program will start sorting the subfile $L(k), \ldots, L(i - 1)$ and later must sort the subfile $L(i + 1), \ldots, L(m)$; so only $i + 1$ and m need be put on the stack before sorting the first subfile and nothing need be saved before sorting the second one. The extra stack bookkeeping is avoided by manipulating the stack in the program instead of leaving it to the compiler for a recursive language. A nonrecursive program may look something like this:

Algorithm 2.5 NONRECURSIVE QUICKSORT

Input: L, the array of keys, and $n \geq 0$, the number of keys.

Output: L with keys in nondecreasing order.

1.   PUSH $([1, n])$
2.   while stack is not empty do
3.   POP $([k, m])$
4.   while $k < m$ do
5.   SPLIT (k, m, i)
6.   PUSH $([i + 1, m])$
7. $m \leftarrow i - 1$
          end
      end

4.   Let b be the amount of space needed on the stack to store the borders of one subfile. SPLIT may split a file into one very large section and one very small one. If

this happens repeatedly and the smaller section is the one whose borders are stacked for later processing, the stack will need bn locations. On the other hand, if we stack the borders of the larger section and process the smaller one, the stack will never fill more than $b \log n$ spaces because the smaller subfile, the one we will continue to split, is no more than half the size of the subfile from which it was obtained. Thus to keep the stack small, $m - i$ and $i - k$ would be compared and the borders of the larger subfile would be stacked while the smaller is sorted. Note that this change will save space at the expense of extra time spent on bookkeeping, that is, on comparing the sizes of subfiles. If very large files are being sorted, this space savings may be necessary.

5.   The four modifications described above have been discussed independently, but they are compatible and can be combined in one program. A good program will use them all.

REMARKS

In practice, QUICKSORT programs run quite fast on the average for large n, and they are widely used. In the worst case QUICKSORT behaves poorly. BUBBLESORT, MAXSORT (see Exercise 2.1), and QUICKSORT are all $\Theta(n^2)$ in the worst case. Are there sorting algorithms that do better, or can we establish a worst-case lower bound of $\Theta(n^2)$? How many comparisons are necessary to sort n keys? These questions will be answered in the next two sections.

2.4   *Lower Bounds for Sorting by Comparison of Keys*

In this section we derive lower bounds for the number of comparisons that must be done in the worst case and on the average by any algorithm that sorts by comparison of keys. To derive the lower bounds we will assume that the keys in the file to be sorted are distinct.

Let n be fixed and suppose that the keys are $x_1, x_2, \ldots, x_n$. We will associate with each algorithm and positive integer n a (binary) decision tree that describes the sequence of comparisons carried out by the algorithm on any input of size n. It is easier to define these decision trees for algorithms that do not have loops and consist only of three kinds of instructions: a compare instruction with a two-way branch, an output instruction (for outputting a permutation on n letters), and a STOP instruction. Let SORT be any algorithm that sorts by comparison of keys. There is an algorithm of the special type just described which is equivalent to SORT in the sense that for each input of size n it does the same comparisons as SORT, in the same order, and outputs the permutation by which SORT rearranges the keys. The decision tree for SORT is the decision tree for this loopless algorithm. It is defined inductively by associating a tree to each compare and output instruction as follows. The tree associated with an output instruction consists of one node labeled with the permutation that is output.

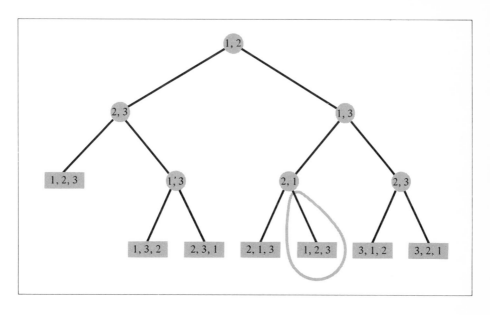

Figure 2.3
Decision tree for BUBBLESORT. The tree shown describes the comparisons in
Algorithm 2.1. The leaves are labeled $\alpha(1)$, $\alpha(2)$, $\alpha(3)$ where $\alpha(i)$ is the correct position
of the key x_i. Note that the circled branch can never be followed; therefore the node
(2,1) can be removed from the tree and replaced by the output node 2,1,3. The resulting
tree is the decision tree for Algorithm 2.2.

The tree associated with an instruction that compares keys x_i and x_j consists of a
root labeled (i,j), a left subtree which is the tree associated with the instruction
executed next if $x_i < x_j$, and a right subtree which is the tree associated with the
instruction done next if $x_i > x_j$. The decision tree for the entire algorithm (and for
SORT) is the tree associated with the first compare instruction it executes. An
example of a decision tree for $n = 3$ is shown in Fig. 2.3.

The action of SORT on a particular input file corresponds to following one path
in its decision tree from the root to a leaf. The tree must have at least $n!$ leaves because
there are $n!$ ways in which the keys may be permuted. Since the unique path followed
for each input depends only on the ordering of the keys and not on their particular
values, exactly $n!$ leaves can be reached from the root by actually executing SORT.
We will assume that any paths in the tree that are never followed are removed. We also
assume that comparison nodes with only one child are removed and replaced by the
child, and that this "pruning" is repeated until all internal nodes have degree 2. The
pruned tree represents an algorithm that is at least as efficient as the original one, so
the lower bounds we derive using decision trees with exactly $n!$ leaves and all internal
nodes of degree 2 will be valid lower bounds for all algorithms that sort by comparison
of keys. From now on we assume SORT is described by such a tree.

The number of comparisons done by SORT on a particular input file is the number of internal nodes on the path followed for that input. Thus the number of comparisons done in the worst case is the number of internal nodes on the longest path, that is, the depth of the tree. The average number of comparisons done is the average of the lengths of all paths from the root to a leaf. (For example, it is clear from Fig. 2.3 that for $n = 3$, BUBBLESORT (Algorithm 2.2) does three comparisons in the worst case and $2\frac{2}{3}$ on the average.) To get the worst-case lower bound we derive a lower bound for the depth of a binary tree in terms of the number of leaves, since the only quantitative information we have about decision trees is the number of leaves.

Lemma 2.1 Let ℓ be the number of leaves in a binary tree and let d be its depth. Then $\ell \leqslant 2^d$.

Proof. A straightforward induction on d.   □

Lemma 2.2 Let ℓ and d be as in Lemma 2.1. Then $d \geqslant \lceil \log \ell \rceil$.

Proof. Taking logs of both sides of the inequality in Lemma 2.1 gives $\log \ell \leqslant d$. Since d is an integer, $d \geqslant \lceil \log \ell \rceil$.   □

Lemma 2.3 For a given n, the decision tree for any algorithm that sorts by comparison of keys has depth at least $\lceil \log n! \rceil$.

Proof. Let $\ell = n!$ in Lemma 2.2.   □

So the number of comparisons needed to sort in the worst case is at least $\lceil \log n! \rceil$, but how does this number compare with $[n(n-1)]/2$, the number of comparisons done by BUBBLESORT and QUICKSORT in the worst case? There are several ways to estimate or get a lower bound for $\log n!$. Perhaps the simplest is to observe that

$$n! \geqslant n(n-1)\cdots\left(\left\lceil\frac{n}{2}\right\rceil\right) \geqslant \left(\frac{n}{2}\right)^{n/2}$$

so

$$\log n! \geqslant \frac{n}{2}\log\frac{n}{2},$$

which is $\Theta(n \log n)$. To get a closer lower bound we use the fact that

$$\log n! = \sum_{j=1}^{n} \log j.$$

This sum is equal to the sum of the areas of the rectangles shown in the graph of the function $y = \log x$ in Fig. 2.4. The sum of the areas of the rectangles is greater than the area under the graph of $\log x$ between 1 and n. So*

$$\log n! = \sum_{j=1}^{n} \log j \geqslant \int_{1}^{n} \log x \, dx.$$

---

* The reader unfamiliar with integrals may look at Fig. 2.10 where $\sum_{j=1}^{n-1} \lfloor \log j \rfloor$, a lower bound for $\sum_{j=1}^{n} \log j$, is also interpreted as an area and evaluated, but without the use of calculus.

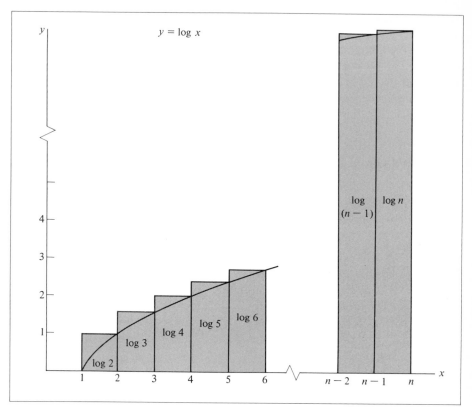

Figure 2.4
$$\sum_{j=2}^{n} \log j > \int_{1}^{n} \log x \, dx.$$

Since $\log x = \log e \cdot \ln x$, where ln is the natural logarithm, we have

$$\int_{1}^{n} \log x \, dx = \log e \int_{1}^{n} \ln x \, dx = \log e \, [x \ln x - x]_{1}^{n}$$

$$= \log e \, (n \ln n - n + 1)$$

$$= n \log n - n \log e + \log e \geq n \log n - n \log e$$

Since $\log e \approx 1.44$, the depth of the decision tree is at least $\lceil n \log n - 1.5 \, n \rceil$.

Theorem 2.4 Any algorithm to sort n items by comparisons of keys must do at least $\lceil \log n! \rceil$, or approximately $\lceil n \log n - 1.5n \rceil$, such comparisons in the worst case.

Theorem 2.4 does not tell us that sorting *can* be done with $\Theta(n \log n)$ comparisons; it says only that *at least* $\Theta(n \log n)$ comparisons are necessary. But it is encouraging that our work did not yield a lower bound of $\Theta(n^2)$. Is there an algorithm that sorts in $\Theta(n \log n)$ comparisons? Or should we try to establish a larger lower bound? Consider the case where $n = 5$. BUBBLESORT and QUICKSORT do $[n(n-1)]/2$, or 10, comparisons in the worst case. If the lower bound in Theorem 2.4 is the best possible, then five items could be sorted with $\lceil \log 5! \rceil = \lceil \log 120 \rceil = 7$ comparisons. The reader should take some time now to work on Exercise 2.15, and notice that the exercise does not specify what the optimal number of comparisons is. It is somewhere between 7 and 10. In the following section we will continue the search for an algorithm that is better than $\Theta(n^2)$ in the worst case.

Now we will compute a lower bound for the average behavior of algorithms that sort by comparison. The *external path length* of a tree is the sum of the lengths of all paths from the root to a leaf; it will be denoted by *epl*. Recall that every node in a decision tree has degree 0 or 2. Binary trees with this property are called *b-trees*.

Lemma 2.5 Among b-trees with ℓ leaves, the *epl* is minimized if all of the leaves are on at most two adjacent levels.

Proof. Suppose we have a b-tree with depth d that has a leaf X at level k, where $k \leqslant d - 2$. We will exhibit a b-tree with the same number of leaves and lower *epl*. Choose a node Y at level $d - 1$ that is not a leaf, remove its children, and attach two children to X. See Fig. 2.5 for an illustration. The total number of leaves has not changed. The *epl* has been decreased by $2d + k$, because the paths to the children of

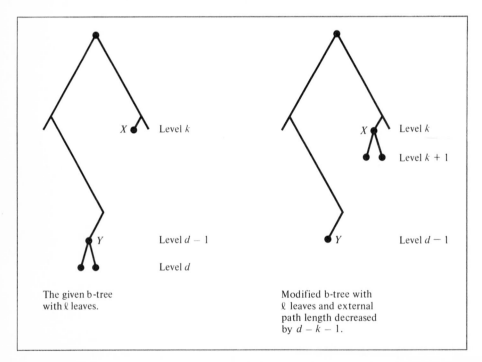

The given b-tree with ℓ leaves.

Modified b-tree with ℓ leaves and external path length decreased by $d - k - 1$.

Figure 2.5
Decreasing external path length.

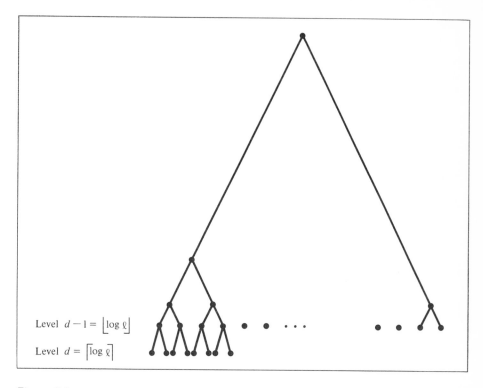

Level $d-1 = \lfloor \log \ell \rfloor$

Level $d = \lceil \log \ell \rceil$

Figure 2.6
Computing external path length for Lemma 2.6. $\ell = 2^{d-1}$ + the number of nodes at level $d-1$ that are not leaves.

Y and the path to X are no longer counted, and increased by $2(k+1)+d-1 = 2k+d+1$, the sum of the lengths of the paths to Y and the new children of X. There is a net decrease in the *epl* of $2d+k-(2k+d+1)=d-k-1>0$ since $k \leqslant d-2$. $\qquad \square$

Lemma 2.6 The minimum *epl* for b-trees with ℓ leaves is $\ell \lfloor \log \ell \rfloor + 2(\ell - 2^{\lfloor \log \ell \rfloor})$.

Proof. If ℓ is a power of 2, all the leaves are at level $\log \ell$. (This statement depends on both the facts that the tree is a b-tree and that all leaves are on at most two levels. The reader should verify it.) The *epl* is $\ell \log \ell$, which is the value of the expression in the lemma in this case.

If ℓ isn't a power of 2, the depth of the tree is $d = \lceil \log \ell \rceil$ and all the leaves are at levels $d-1$ and d. The sum of the path lengths (for all leaves) down to level $d-1$ is $\ell(d-1)$. For each leaf at level d, 1 must be added to get the total *epl*. The number of leaves at level d is $2(\ell - 2^{d-1})$, since for each node at level $d-1$ that is not a leaf, there are two leaves at level d. (See Fig. 2.6.) Thus the sum is $\ell(d-1) + 2(\ell - 2^{d-1}) = \ell \lfloor \log \ell \rfloor + 2(\ell - 2^{\lfloor \log \ell \rfloor})$. $\qquad \square$

Theorem 2.7 The average number of comparisons done by an algorithm to sort n items by comparison of keys is at least $\lfloor \log n! \rfloor \approx \lfloor n \log n - 1.5n \rfloor$.

Proof. The average number of comparisons is at least the minimum average path length for a decision tree with $n!$ leaves; in other words, at least

$$\frac{n! \lfloor \log n! \rfloor + 2(n! - 2^{\lfloor \log n! \rfloor})}{n!} = \lfloor \log n! \rfloor + \epsilon,$$

where $0 \leqslant \epsilon < 1$ since $n! - 2^{\lfloor \log n! \rfloor}$ is always less than $n!/2$. $\qquad\square$

We already have an algorithm, QUICKSORT, which does $\Theta(n \log n)$ comparisons on the average. Theorem 2.7 says there is no algorithm that is substantially better.

2.5   *HEAPSORT*

The HEAPSORT algorithm uses a data structure called a heap, which is a binary tree with some special properties. The definition of a heap includes a description of the structure and a condition on the data in the nodes. Informally, a heap structure is a complete binary tree with some of the rightmost leaves removed. See Fig. 2.7 for

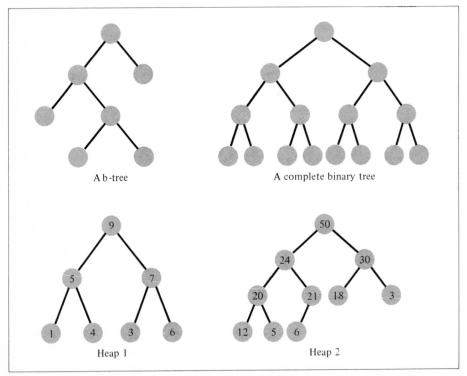

A b-tree

A complete binary tree

Heap 1

Heap 2

Figure 2.7
B-trees, complete binary trees, and heaps.

illustrations. Let S be a set of keys with a linear ordering and let T be a binary tree with depth d whose nodes contain elements of S. T is a *heap* if and only if it satisfies the following conditions:

1. All internal nodes (with one possible exception) have degree 2, and at level $d - 1$ the leaves are all to the right of the internal nodes. The rightmost internal node at level $d - 1$ may have degree 1 (with no right child).

2. The key at any node is greater than or equal to the keys at each of its children (if it has any).

We will use the term *heap structure* to describe a binary tree that satisfies condition (1). Observe that a complete binary tree is a heap structure. When new nodes are added to a heap, they must be added left to right at the bottom level, and if a node is removed, it must be the rightmost node at the bottom level if the resulting structure is still to be a heap. Note that the root must contain the largest key in the heap.

DELETION FROM A HEAP

Deletion from a heap means removing the key at the root, the largest key in the heap, and rearranging the nodes so that the heap properties are still satisfied. Structurally, the node to be removed is the rightmost leaf at the bottom level. The key, say K, from that leaf must be put elsewhere. The only vacant node is the root, so we begin there and let K filter down to its correct position. At its final position, K must be greater than or equal to each of its children, so at each step K is compared to the larger of the children of the currently vacant node. If K is larger (or equal) it can be inserted; otherwise the larger child is moved up to the vacant node and the process is repeated. The deletion algorithm assumes that there are at least two nodes in the heap. The algorithm is illustrated in Fig. 2.8.

Algorithm 2.6 DELETE FROM A HEAP

Input: A heap and ROOT, a pointer to its root.

Output: A heap with one less node containing all the keys from the input heap except the one that was at the root.

1. Copy KEY(ROOT) elsewhere.
2. Let K be the key in the rightmost leaf at the bottom level. Remove that leaf from the heap structure.
3. $P \leftarrow$ ROOT
4. **while** P is not a leaf **do**
5.   Set M to point to the child of P with the larger key.
6.   **if** $K < $ KEY(M) **then do**
7.   KEY(P) $\leftarrow$ KEY(M)
8. $P \leftarrow M$
          end
9.   **else exit**
   end
10. KEY(P) $\leftarrow K$

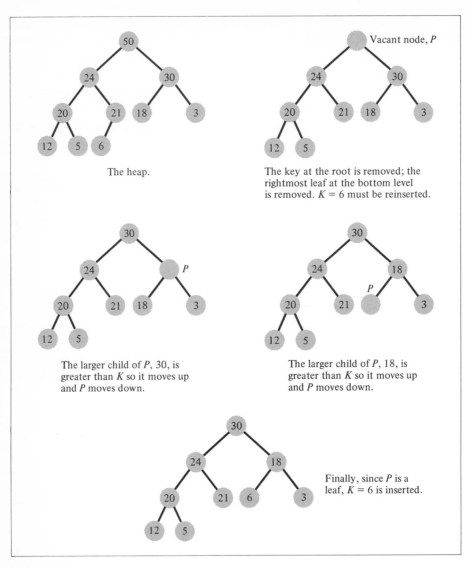

Figure 2.8
Deletions from a heap.

Lemma 2.8 The heap deletion algorithm does $2d$ comparisons of keys in the worst case on a heap with depth d after the deletion.

Proof. Two comparisons of keys are done at lines 5 and 6. The maximum number of times these instructions are executed is the depth of the heap (after the deletion).   □

Clearly we could sort a file by arranging the keys in a heap, and then removing them one at a time in nonincreasing order using the deletion algorithm. The file can be filled from the end with keys removed from the heap to get them in nondecreasing order. Thus we need an algorithm to construct a heap.

HEAP CONSTRUCTION

The scheme for arranging the keys in a heap is to make a large heap from two smaller ones. Suppose P points to the root of a heap structure [Property (1) of the definition] in which each of the subtrees is a heap [Properties (1) and (2)]. The key at the root may not be the largest key, so to make the whole thing into a heap some rearranging must be done. The steps to be carried out are those in lines 4–10 of the algorithm to delete a node from a heap where K is the key at the root. Thus we define the following.

Algorithm 2.7   FIXHEAP(P)

Input: A heap structure in which the two subtrees are heaps, and P, a pointer to the root.

Output: The same tree with keys rearranged to make it a heap.

1. $K \leftarrow \text{KEY}(P)$
2.   Lines 4–10 of Algorithm 2.6.

If all the keys to be sorted are put into a heap structure in an arbitrary order, FIXHEAP can be used to build a heap if we can find some little heaps to start with. The observation that each leaf is a heap leads to the following algorithm.

Algorithm 2.8 HEAP CONSTRUCTION

Input: A heap structure [Property (1)] with keys in arbitrary nodes.

Output: The same structure arranged as a heap.

1.   Let d be the depth of the heap.
2.   **for** $\ell \leftarrow d - 1$ **to** 0 **by** -1 **do**
3.   **for** each non-leaf P at level ℓ
4.   **do** FIXHEAP (P) **end**
     end

    The HEAPSORT algorithm will consist of using Algorithm 2.8, and then Algorithm 2.6 repeatedly. It is not yet clear that this is a good algorithm; it seems to require extra space, and the implementation of some steps in the algorithms (e.g., line 3 in Algorithm 2.8) may be complicated. We will examine the implementation before writing the HEAPSORT algorithm and doing the worst-case analysis.

IMPLEMENTATION OF A HEAP AND THE HEAPSORT ALGORITHM

Binary trees are generally implemented as linked structures; that is, each node contains pointers to the roots of its subtrees. Setting up and using such a structure requires extra bookkeeping and extra space for the pointers. However, we store and use a heap without any pointers at all. Because in a heap there are no nodes at, say, level ℓ unless

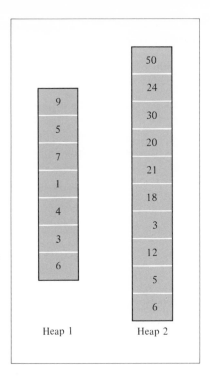

Figure 2.9
Storage of the heaps in Fig. 2.7.

level $\ell - 1$ is completely filled, a heap can be stored efficiently in an array in such a way that accessing a child of a node is quite easy. The nodes are stored in the array from left to right and level by level (beginning with the root). Figure 2.9 shows the storage arrangement for the heaps in Fig. 2.7. Suppose the index i of a node is given. Then a counting argument can be used to show that $\text{LCHILD}(i) = 2i$ and $\text{RCHILD}(i) = 2i + 1$ (see Exercise 2.17).

The startling feature of **HEAPSORT** is that the whole sorting procedure can be carried out in place; the small heaps built during the construction phase, and, later, the heap and the deleted keys, can occupy the array L that originally contained the unsorted file. During the deletion phase, when the heap contains i keys, it will occupy the first i locations in the array. Just one variable, i, is needed to mark the end of the heap. The reader should be able to determine how to test whether a node is a leaf (for FIXHEAP) and how to implement various other details.

Algorithm 2.9 HEAPSORT

Input: L, an unsorted array, and $n \geqslant 1$, the number of keys.

Output: L with keys in nondecreasing order.

1. Algorithm 2.8: HEAP CONSTRUCTION
2. **for** $i \leftarrow n$ **to** 2 **by** -1 **do**

>   Delete from the heap with i nodes (the first i cells of L), using FIXHEAP with appropriate initialization, and store the deleted key in $L(i)$.

end

WORST-CASE ANALYSIS

Let $d = \lfloor \log n \rfloor$, the depth of a heap with n nodes. By Lemma 2.8 the number of comparisons done by FIXHEAP(P) in the worst case is twice the depth of the tree rooted at P, which is $d - $ (the level of P). Thus the number of comparisons done by the heap construction is at most

$$\sum_{\ell=0}^{d-1} 2(d - \ell)(\text{the number of nodes at level } \ell) \;=\; 2 \sum_{\ell=0}^{d-1} (d - \ell)2^{\ell}$$

$$= 2d \sum_{\ell=0}^{d-1} 2^{\ell} - 2 \sum_{\ell=0}^{d-1} \ell 2^{\ell}$$

$$= 2d(2^d - 1) - 4 \sum_{\ell=1}^{d-1} \ell 2^{\ell-1}$$

$$= 2d(2^d - 1) - 4[d2^{d-1} - 2^d + 1]$$

$$= 2^{d+2} - 2d - 4.$$

(The formula for $\sum_{\ell=1}^{d-1} \ell 2^{\ell-1}$ was derived in Section 1.4.) In terms of n, the number of comparisons done to construct the heap is at most $cn - 2\lfloor \log n \rfloor$, where $2 \leqslant c < 4$; it is linear in n.

By Lemma 2.8 again, the number of comparisons done to delete a node from a heap with i nodes remaining is at most $2\lfloor \log i \rfloor$, so the total for all the deletions is at most $2\sum_{i=1}^{n-1} \lfloor \log i \rfloor$. To evaluate the sum $\sum_{i=1}^{n-1} \lfloor \log i \rfloor$ we use Fig. 2.10, which illustrates the case where $n = 10$. $\sum_{i=1}^{n-1} \lfloor \log i \rfloor$ equals the sum of the areas of the rectangles shown in the figure; it is

$$\sum_{j=1}^{\lfloor \log n \rfloor -1} j2^j + \lfloor \log n \rfloor (n - 2^{\lfloor \log n \rfloor}),$$

where the summation term includes the areas of all the complete rectangles (height j and width 2^j) and the second term is the area of the last, incomplete rectangle with height $\lfloor \log n \rfloor$ and width $n - 2^{\lfloor \log n \rfloor}$. The total area is

$$2\left[\lfloor \log n \rfloor \, 2^{\lfloor \log n \rfloor -1} - 2^{\lfloor \log n \rfloor} + 1 \right] + \lfloor \log n \rfloor (n - 2^{\lfloor \log n \rfloor})$$

$$= n\lfloor \log n \rfloor - 2^{\lfloor \log n \rfloor +1} + 2.$$

The following theorem sums up our results.

Theorem 2.9 HEAPSORT does at most $2n \lfloor \log n \rfloor$ comparisons of keys in the worst case. It is an $\Theta(n \log n)$ sorting algorithm.

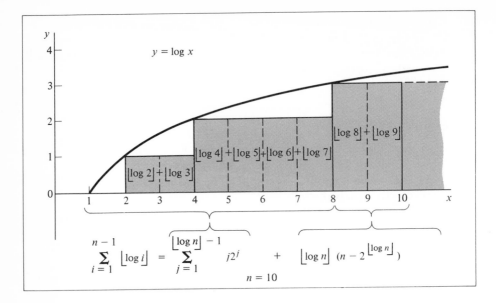

Figure 2.10

Proof. Heap construction requires at most $2^{\lfloor \log n \rfloor + 2} - 2\lfloor \log n \rfloor - 4$ comparisons, and the deletions do at most $2[n\lfloor \log n \rfloor - 2^{\lfloor \log n \rfloor + 1} + 2]$. The total is $2n\lfloor \log n \rfloor$. $\square$

REMARKS

HEAPSORT does $\Theta(n \log n)$ comparisons on the average as well as in the worst case. Table 2.1 sums up the results of the analysis of the behavior of the three sorting algorithms that have been discussed so far. There are algorithms that do fewer comparisons in the worst case than HEAPSORT. The lower bound obtained in Section 2.4 is quite good. It is known to be exact for some values of n; that is, $\lceil \log n! \rceil$ com-

Table 2.1

Results of analysis of three sorting algorithms (entries are approximate).

Algorithm	Worst case	Average	Space usage
BUBBLESORT	$\dfrac{n^2}{2}$	$\dfrac{n^2}{4}$	In place
QUICKSORT	$\dfrac{n^2}{2}$	$\Theta(n \log n)$	Extra space proportional to $\log n$
HEAPSORT	$2n\lfloor \log n \rfloor$	$\Theta(n \log n)$	In place

parisons are sufficient to sort, for some values of n. It is also known that $\lceil \log n! \rceil$ comparisons are not sufficient for all n. For example $\lceil \log 12! \rceil \doteq 29$, but it has been proved that 30 comparisons are necessary (and sufficient) to sort 12 items in the worst case. See the notes and references at the end of the chapter for references on sorting algorithms whose worst-case behavior is close to the lower bound.

2.6   *SHELLSORT*

The SHELLSORT algorithm is included in this chapter because the technique it uses is interesting, and it is easy to program and runs fairly quickly. The analysis of the algorithm is, however, very difficult and incomplete. SHELLSORT sorts a file L with n keys by successively sorting larger and larger subfiles whose entries are intermingled in the whole file. The subfiles to be sorted are determined by a sequence, h_t, $h_{t-1}, \ldots, h_1$, of parameters called increments. The final increment, h_1, is 1. Suppose for example that the first increment, h_t, is 6. Then the file is divided into six subfiles, each beginning with one of the first six keys and consisting of every sixth key from there on. After these subfiles are sorted, the next increment h_{t-1} is used to separate the file again into subfiles, this time with entries h_{t-1} elements apart, and again subfiles are sorted. The process is repeated for each increment after which, since $h_1 = 1$, the entire file will be sorted. Figure 2.11 illustrates the action of this method on a small file.

The informal description given of the method used by SHELLSORT should prompt a number of questions from the reader. What algorithm should be used to sort the subfiles? Considering that the last increment is 1 and the entire file is sorted on the last pass, is SHELLSORT any more efficient than the algorithm used to sort the subfiles? Can the algorithm be written to minimize all the bookkeeping that seems to be needed to control the sorting of all the subfiles? What increments should be used?

We tackle the first two questions first. As the example in Fig. 2.11 shows, when the last few passes are made using small increments, few keys will be out of order because of all the work that was done in earlier passes. So SHELLSORT may be efficient, and indeed would only be efficient, if the method used to sort subfiles is one that does very little work if the file is already sorted or nearly sorted. The particular method used is called INSERTION SORT. For each j between 2 and n, INSERTION SORT inserts $L(j)$ into the already sorted segment of the file, $L(1), \ldots, L(j-1)$, by comparing $L(j)$ to these keys (from right to left) and moving larger keys to the right until the proper spot for $L(j)$ is found. The algorithm is given below.

Algorithm 2.10 INSERTION SORT

Input: L, an unsorted array, and $n \geqslant 1$, the number of keys.

Output: L with keys in nondecreasing order.

1.   **for** $j \leftarrow 2$ **to** n **do**
2. $KEY \leftarrow L(j); i \leftarrow j - 1$

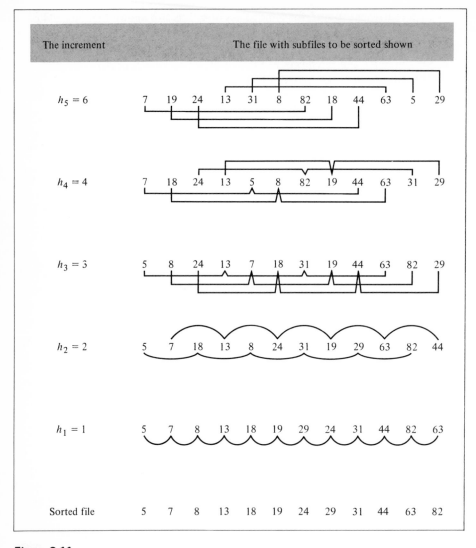

Figure 2.11
SHELLSORT. (Note that only two pairs of keys must be interchanged on the final pass.)

3.   **while** $i > 0$ and $L(i) > \text{KEY}$ **do**
4. $L(i + 1) \leftarrow L(i); i \leftarrow i - 1$
            end
5. $L(i + 1) \leftarrow \text{KEY}$
        end

INSERTION SORT is even simpler than BUBBLESORT, and it is as inefficient.

In the worst case KEY is compared to $L(i)$ in line 3 for each i between 1 and $j-1$ and for each j, so the total number of comparisons would be

$$\sum_{j=2}^{n} (j-1) = \frac{n(n-1)}{2}.$$

(By now the reader should be convinced that it is easy to write sorting algorithms whose worst-case behavior is $\Theta(n^2)$.) The average behavior of INSERTION SORT is also $\Theta(n^2)$ because it removes at most one inversion after each comparison. The important characteristics of INSERTION SORT for the purposes of SHELLSORT are that it does little work when the file is nearly sorted (only $n-1$ comparisons if it is completely sorted), it is extremely simple to program, and there is very little overhead for bookkeeping.

Now suppose SHELLSORT is using an increment h and is to sort h subfiles, each containing approximately n/h entries. If each subfile is to be completely sorted before any work is begun on the next, the algorithm would need a counter to keep track of which subfiles have been sorted and which remain to be done. This bookkeeping is avoided by having the algorithm make one pass through the entire file for each increment intermingling its work on all the subfiles. Since consecutive keys of a subfile are h cells apart rather than one apart, "1" is replaced by h in INSERTION SORT (lines 3–7 of the following algorithm).

Algorithm 2.11 SHELLSORT

Input: L, an unsorted array; n, the number of keys; and increments $h_t, h_{t-1}, \ldots, h_1$, where $h_1 = 1$.

Output: L with keys in nondecreasing order.

1.   **for** $s \leftarrow t$ **to** 1 **by** -1 [s indexes the increments] **do**
2. $h \leftarrow h_s$
3.   **for** $j \leftarrow h+1$ **to** n [j begins at the second key of the first file] **do**
4.   KEY $\leftarrow L(j); i \leftarrow j-h$
5.   **while** $i > 0$ and $L(i) >$ KEY **do**
6. $L(i+h) \leftarrow L(i); i \leftarrow i-h$
              end
7. $L(i+h) \leftarrow$ KEY
         end
     end

Observe that although after each comparison INSERTION SORT removes at most one inversion from the file it is sorting, the way it is used in SHELLSORT causes keys to be moved across many others with the chance of eliminating many inversions for each comparison. Thus there is a possibility that the average behavior of SHELLSORT is better than $\Theta(n^2)$. The efficiency of SHELLSORT stems from the fact that sorting with one increment, say k, will not undo any of the work done previously when a different increment, say h, was used. More precisely, we say that a list is h-ordered if $L(i) \leqslant L(i+h)$ for $1 \leqslant i \leqslant n-h$, or in other words, if all the subfiles consisting of every hth key are sorted. To k-sort a file means to sort subfiles using increment k.

Theorem 2.10 If an h-ordered file is k-sorted, the file will still be h-ordered.

Proof. See Knuth (1973) for the proof. The reader should examine Fig. 2.11 to see that the theorem is true for the example given there. □

ANALYSIS

The number of comparisons done by SHELLSORT is a function of the sequence of increments used. The analysis is extremely difficult and requires answers to some mathematical problems that have not yet been solved. Therefore, the best possible sequence of increments has not been determined, but some specific cases have been thoroughly studied. One of these is the case where $t = 2$, that is, where exactly two increments, h and 1, are used. It has been shown that the best choice for h is approximately $1.72 \sqrt[3]{n}$, and with this choice the average running time is proportional to $n^{5/3}$. This may seem surprising since using the increment 1 is the same as doing INSERTION SORT, which is $\Theta(n^2)$ on the average; just doing one preliminary pass through the file with increment h lowers the order of the running time. By using more than two increments, the running time can be improved even more.

It is known that if the increments $h_k = 2^k - 1$ for $1 \leqslant k \leqslant \lfloor \log n \rfloor$ are used, the number of comparisons done in the worst case is $O(n^{3/2})$. These increments are very easy to compute in assembly language; each one (after the first) can be obtained from the previous one by shifting one bit to the right. Empirical studies (with values of n as high as 250,000) have shown that another set of increments gives rise to very fast running programs. These are defined by $h_i = (3^i - 1)/2$ for $1 \leqslant i \leqslant t$, where t is chosen as the smallest integer such that $h_{t+2} \geqslant n$. These increments are easy to compute iteratively. We can find h_t at the beginning of the sort by using the relation $h_{s+1} = 3h_s + 1$ and comparing the results to n. Instead of storing all the increments, they can be recomputed in reverse order during the sort using the relation $h_s = (h_{s+1} - 1)/3$.

It has been proved that if the increments are all integers of the form $2^i 3^j$ which are less than n (used in decreasing order), then the number of comparisons done is $O(n(\log n)^2)$. The worst-case running times for the other sets of increments are known or expected to be of higher order. However, because of the large number of integers of the form $2^i 3^j$, more passes through the file, hence more bookkeeping, will be done with these increments than with others. Therefore, they are not particularly useful unless n is fairly large.

2.7   *Bucket Sorts*

For the sorting algorithms in Sections 2.2 through 2.6, only one assumption was made about the keys: that they are elements of a linearly ordered set. The basic operation of the algorithms is a comparison of two keys to determine which is larger. If we make more assumptions about the keys, we can consider algorithms that perform other operations on them. Suppose the keys are names and are printed on cards, one name per card. To alphabetize the cards by hand we might first separate them into 26 different piles according to the first letter of the name, or fewer piles with several letters in each; alphabetize the cards in each pile by some other method perhaps similar to INSERTION SORT; and finally combine the sorted piles. If the keys are all five-digit decimal integers we might separate them into ten piles according to the first digit. If they are integers between 1 and m, for some m, we might make a pile for each of the k intervals $[1, m/k]$, $[m/k + 1, 2m/k]$, and so on. In each of these examples the keys are distributed into different piles as a result of examining individual letters or digits in a key or by comparing keys to predetermined values. Then the piles are sorted individually and recombined. Algorithms that sort by such methods are called "bucket sorts" or "algorithms that sort by distribution." These algorithms are not in the class of algorithms previously considered because to use them we must know something about either the structure or the range of the keys.

How fast are bucket sorts? To answer this question we must decide on a measure of work. A bucket sort has three phases, which we may call distribution, sorting buckets, and combining buckets. The type of work done in each phase is different. Suppose there are k buckets. During the distribution phase, the algorithm examines each key once (either examining a particular field of bits or comparing the key to at most k preset values). Then it does some work to indicate in which bucket the key belongs. This might involve copying the key or setting some indexes or pointers. Without knowing more about the implementation details of a specific algorithm, we can conclude only that the amount of work done in the first phase is proportional to the number of keys, n. Suppose that to sort the buckets we use an algorithm that sorts by comparison of keys doing, say, $S(m)$ comparisons for a bucket with m keys. Let n_i be the number of keys in the ith bucket. The algorithm does $\Sigma_{i=1}^{k} S(n_i)$ comparisons during the second phase. The third phase may require, at worst, that all of the keys be copied from the buckets into one file; the amount of work done is $O(n)$. Thus, most of the work is done while sorting buckets. Suppose $S(m)$ is $\Theta(m \log m)$. Then if the keys are uniformly distributed among the buckets, the algorithm does roughly $ck(n/k) \log (n/k) = cn \log (n/k)$ comparisons of keys in the second phase, where c depends on the sorting algorithm used in the buckets. Increasing k, the number of buckets, decreases the number of comparisons done. If we choose $k = n/10$, then $n \log 10$ comparisons would be done and the running time of the bucket sort would be linear in n, assuming that the keys are evenly distributed and that the timing of the first phase doesn't depend on k. However, in the worst case, all of the keys will go into one bucket and the entire file will be sorted in the second phase, making all of the work of the first and last phase completely unnecessary. Thus in the worst case a bucket sort would be very inefficient. If the distribution of the keys is known in advance, the range of keys to go into each bucket can be adjusted so that all buckets receive an approximately equal number of keys.

The amount of space needed by a bucket sort depends on how the buckets are stored. Most of the sorting algorithms studied in the previous sections assume that the keys is known in advance, the range of keys to go into each bucket can be accessed in a fixed, short amount of time. If every bucket is to consist of a set of sequential locations, then each must be allocated enough space to hold the maximum number of keys that might belong in one bucket, and that is n. Thus kn locations would be allocated to sort n keys. As the number of buckets increases, the speed of the algorithm would increase but so would the amount of space used. Linked lists are a very appropriate alternative to sequential storage in such a problem. Space for n keys plus links and a listhead for each bucket would be needed; distributing keys among the buckets would require adjusting pointers. But then how would the keys in each bucket be sorted? QUICKSORT, HEAPSORT, and SHELLSORT, the faster algorithms discussed, cannot sort linked files. If the number of buckets is large, the number of keys in each will generally be small and a slower algorithm could be used. BUBBLESORT can be easily modified to sort keys in a simply linked list. With approximately n/k keys per bucket, BUBBLESORT will do approximately $n^2/4k^2$ comparisons on the average for each bucket, or $n^2/4k$ comparisons in all. Here again, as k increases, so does the speed, but since there are k listheads for the buckets the amount of space used increases. However, the total amount of space used for linked buckets is less by roughly a factor of k than the amount needed for buckets stored in sequential locations.

The reader might wonder why we don't use a bucket sort algorithm recursively to create smaller and smaller buckets. There are several reasons. The bookkeeping would quickly get out of hand; pointers indicating where the various buckets begin and information needed to recombine the keys into one file would have to be stacked and unstacked often. Due to the amount of bookkeeping necessary for each recursive call, the algorithm should not count on ultimately having only one key per bucket, so another sorting algorithm will be used anyway to sort small buckets. Thus if a fairly large number of buckets is used in the first place, there is little to gain and a lot to lose by bucket sorting recursively. However, although recursively distributing keys into buckets is not efficient, something useful can be salvaged from this idea.

RADIX SORTING

Suppose that the keys are five-digit numbers. A recursive algorithm would first distribute the keys among ten buckets according to the leftmost, or most significant, digit and then distribute the keys in each bucket among ten more buckets according to the next most significant digit, and so on. The buckets could not be coalesced until they are completely sorted, hence the large amount of messy bookkeeping. It is startling that if the keys are distributed into buckets first according to their *least significant* digits (or bits, letters, or fields), then the buckets can be coalesced in order before distributing on the next (to the left) digit. Observe that the problem of sorting the buckets has been completely eliminated. If there are, say, five digits in a key then the algorithm distributes keys into buckets and then coalesces the buckets five times, distributing on each digit position in turn, right to left, as illustrated in Fig. 2.12. Does this always work? On the final pass when two keys are put into the same bucket because they both start with, say, 9, what ensures that they are in the proper order relative to each other? In Fig. 2.12, the keys 90283 and 90583 differ in the third digit only and are put in the same bucket in each pass except the third. After the third pass, so long as the buckets are coalesced in order and the relative order of two keys placed in the same bucket is not changed, these keys remain in proper order relative to each other. In general, if the leftmost digit position in which two keys differ is the ith position (from the right), they will be in the proper order relative to each other after the ith pass.

This sorting method is called *radix sorting* and is used by card-sorting machines. The machine does the distribution step; the operator collects the piles after each pass and combines them into one for the next pass. In the algorithm below the distribution into buckets is controlled by a particular bit field in the key. For example, if the keys are 32-bit positive integers, the algorithm could use 16 buckets and distribute the keys among them according to four-bit fields extracted from the keys, beginning with the low-order four bits. The number of buckets and the number of fields are constants of the algorithm; they do not depend on the input size.

The variables and data structure used in RADIX SORT, DISTRIBUTE, and COALESCE are as follows:

FILE A pointer. It is assumed that the unsorted file is given as a linked list with FILE pointing to the first node. Each node has a KEY field and a LINK field.

k The number of buckets.

m The number of fields in a key on which the distribution is done.

BUCKET An array of listheads for the buckets. Each bucket is a linked list.

LAST An array of pointers. LAST(i) will point to the last key in the ith bucket to facilitate insertion of keys at the end of a bucket to maintain the relative order.

PTR A pointer.

LOC Means the location of a node or variable.

The data structure is illustrated in Fig. 2.13 for the third pass of the example in Fig. 2.12.

Algorithm 2.12 RADIX SORT

Input: FILE, k, and m as described above.

Output: The sorted file.

Unsorted file	48081	97342	90287	90583	53202	65215	78397	48001	00972	65315	41983	90283	81664	38107
First pass	48081	48001	97342	53202	00972	90583	41983	90283	81664	65215	65315	90287	78397	38107
	Bucket 1		Bucket 2			Bucket 3			Bucket 4	Bucket 5		Bucket 7		
Second pass	48001	53202	38107	65215	65315	97342	81664	00972	48081	90583	41983	90283	90287	78397
	Bucket 0			Bucket 1		Bucket 4	Bucket 6	Bucket 7	Bucket 8					Bucket 9
Third pass	48001	48081	38107	53202	65215	90283	90287	65315	97342	78397	90583	81664	00972	41983
	Bucket 0		Bucket 1	Bucket 2				Bucket 3			Bucket 5	Bucket 6	Bucket 9	
Fourth pass	90283	90287	90583	00972	81664	41983	53202	65215	65315	97342	48001	48081	38107	78397
	Bucket 0				Bucket 1		Bucket 3	Bucket 5		Bucket 7	Bucket 8			
Fifth pass	00972	38107	41983	48001	48081	53202	65215	65315	78397	81664	90283	90287	90583	97342
	Bucket 0	Bucket 3	Bucket 4			Bucket 5	Bucket 6		Bucket 7	Bucket 8	Bucket 9			
Sorted file	00972	38107	41983	48001	48081	53202	65215	65315	78397	81664	90283	90287	90583	97342

Figure 2.12
RADIX SORT.

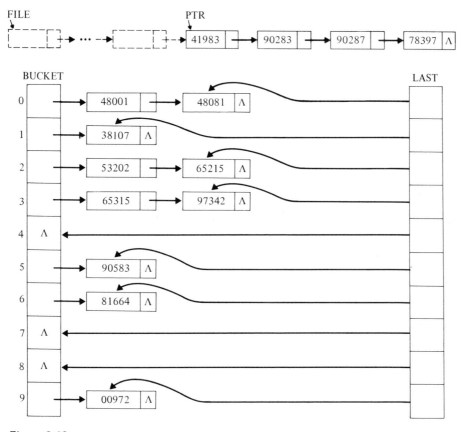

Figure 2.13
The data structure for radix sort during distribution on the third digit.

1. **for** i ← 1 **to** *m* **do**
2.   Set a mask for extracting the *i*th field (from the right, or least significant, end) of a key.
3.   DISTRIBUTE
4.   COALESCE
    end

Algorithm 2.13 DISTRIBUTE

1. **for** *j* ← 1 **to** *k* **do** LAST(*j*) ← LOC(BUCKET(*j*)) **end**
2. PTR ← FILE
3. **while** PTR ≠ Λ **do**
4.   Extract the *i*th field from KEY(PTR) and assign to *j* the appropriate bucket number
5.   LINK(LAST(*j*)) ← PTR; LAST(*j*) ← PTR; PTR ← LINK(PTR)
    end

Algorithm 2.14 COALESCE

1.   PTR ← LOC(FILE)
2.   **for** j ← 1 **to** k **do**
3.   **if** LAST(j) ≠ LOC(BUCKET(j)) [if the jth bucket is not empty]
                then do   LINK(PTR) ← BUCKET(j)
                              PTR ← LAST(j)
                end
      end
6.   LINK(PTR) ← Λ

ANALYSIS

It is reasonable to take the assignment of a key to a bucket (which includes determining which bucket the key belongs in and adjusting the necessary pointers) as the basic operation performed by this algorithm. The running time will be proportional to the number of such assignments. On each pass through the DISTRIBUTE routine, each key is assigned to one bucket. The number of passes is m, the number of fields in each key used for distribution. Thus the number of basic operations is mn. The amount of work done in one pass through the COALESCE routine is proportional to k, the number of buckets, which we assume is small compared to n. Since m and k are fixed, the running time is linear in n. (See Exercise 2.20.)

2.8   *Merging Sorted Lists*

In this section we study the following problem: Given two files A and B sorted in non-decreasing order, merge them to create one sorted file C. As with the sorting algorithms studied in the first six sections of this chapter, the measure of work done by a merge algorithm will be the number of comparisons of keys performed by the algorithm. Let n and m be the number of items in A and B, respectively, and let a_i, b_i, and c_i be the ith entries of A, B, and C, respectively. There is a straightforward solution to the merge problem that works well in some cases. Beginning with the complete files A and B, compare the first remaining keys in A and B and move the smaller one to the next vacant position in C. When A or B is empty, move the items remaining in the other file to C.

Algorithm 2.15 MERGE

Input: A and B, files with keys in nondecreasing order; and n and m, the number of keys in each.

Output: C, a file containing all the keys from A and B in nondecreasing order.

1. $i \leftarrow j \leftarrow k \leftarrow 1$ [i indexes A; j indexes B; k indexes C]
2.   **while** $i \leqslant n$ **and** $j \leqslant m$ **do**
3.   **if** $a_i < b_j$
4.   **then do** $c_k \leftarrow a_i$; $i \leftarrow i + 1$ **end**
5.   **else do** $c_k \leftarrow b_j$; $j \leftarrow j + 1$ **end**
6. $k \leftarrow k + 1$
     end
7.   **if** $i > n$ **then** move $b_j, \ldots, b_m$ to $c_k, \ldots, c_{n+m}$
8.   **else** move $a_i, \ldots, a_n$ to $c_k, \ldots, c_{n+m}$

WORST CASE

Whenever a comparison of a key from A with a key from B is done, at least one key is moved to C and never examined again. After the last comparison, at least two keys are moved to C, the smaller of the two compared and all that remain in the other file. So, at most $n + m - 1$ comparisons are done. The worst case, using all $n + m - 1$ comparisons, occurs when a_n and b_m belong in the last two positions in C. If $n = m$, $2n - 1$ comparisons are done in the worst case. We show next that for this special case (i.e., $n = m$) the algorithm is optimal.

OPTIMALITY WHEN $n = m$

Theorem 2.11 Any algorithm to merge two sorted files, each containing n entries, by comparison of keys does at least $2n - 1$ such comparisons in the worst case.

Proof. Suppose we are given an arbitrary merge algorithm. Keys can be chosen so that the algorithm must compare a_i with b_i, for $1 \leqslant i \leqslant n$, and a_i with b_{i+1}, for $1 \leqslant i \leqslant n - 1$. Specifically, choose keys so that whenever the algorithm compares a_i and b_j, if $i < j$, the result is that $a_i < b_j$, and if $i \geqslant j$, the result is that $a_i > b_j$. Choosing the keys so that $b_1 < a_1 < b_2 < a_2 < \cdots < b_n < a_n$ will satisfy these conditions. However, if for some j, the algorithm never compares a_j and b_j, then choosing keys so that $b_1 < a_1 < b_2 < \cdots < a_{j-1} < a_j < b_j < b_{j+1} < \cdots < b_n < a_n$ will also satisfy these conditions and the algorithm would not be able to determine the correct ordering. Similarly, if for some j, it never compares a_j and b_{j+1}, the arrangement $b_1 < a_1 < \cdots < b_j < b_{j+1} < a_j < \cdots < a_n$ would be consistent with the results of the comparisons done, and again the algorithm could not determine the correct ordering. ☐

SPACE USAGE

It might appear from the way in which the algorithm is written that merging files with a total of N entries requires enough memory locations for $2N$ entries, since entries are copied from files A and B to file C. In some cases, however, the amount of extra

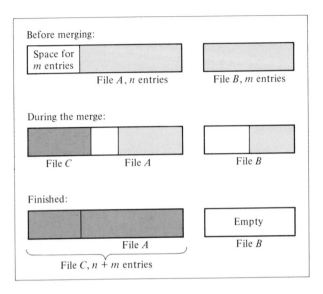

Figure 2.14
Overlapping files for MERGE.

space needed can be decreased. Suppose $n \geqslant m$. Then only m extra locations are needed if file C overlaps file A, as shown in Fig. 2.14. The first m entries moved to C will fill the extra locations. From then on the vacated locations in file A are used. There will always be a gap (i.e., some empty locations) between the end of the merged portion of the file and the remaining entries of file A until all of the entries have been merged. Observe that if this space-saving storage layout is used, line 8 in the merge algorithm can be eliminated because if file B empties before file A, the remaining items in A are in their correct position and do not have to be moved.

BINARY MERGING

It is not always the case that the files being merged are of equal size, and even if $n = m$, after MERGE has moved some of the keys to C, the number of entries remaining in A and B may differ. Thus it is worthwhile to examine the behavior of MERGE when $n \neq m$. Consider the extreme case in which $m = 1$ and n is large. Then MERGE essentially does a sequential search through file A to find the place for b_1. In the worst case n comparisons are done; on the average, $n/2$. But we already know that a binary search can find the place for b_1 much more quickly, doing at most $\lfloor \log n \rfloor + 1$ comparisons. Clearly, even if m is not 1, so long as it is very small compared to n, it would be better to use binary search to find the correct position of each key of B. If the correct positions for the keys of B are fairly uniformly distributed through A, we would expect that b_1 belongs somewhere in the first mth of A, b_2 in the second mth of A, and so forth. If this is the case, a binary search could be done to find the correct position for each entry of B in the appropriate mth of A, instead of searching all of A. Then approximately $m \log (n/m)$ comparisons would be done to merge the lists. Of course it is not always true that b_i belongs in the ith one-mth of A, but it is easy enough to find out. We can begin by comparing b_1 with $a_{\lfloor n/m \rfloor}$. If b_1 is smaller, then a binary search in the first mth of A will find the correct position for b_1. On the other hand, if b_1 is larger (or equal), then as a result of doing just one comparison we know that the entire first mth of A can be moved to C; the problem has been reduced to merging a file of m entries into a file of $n - n/m$. This is basically the strategy of the algorithm presented below; it is called BINARY MERGE. The algorithm differs in several details from the description just given, and we explain these details next. An example follows the algorithm.

To simplify the indexing, BINARY MERGE works from the end of each file rather than from the beginning; that is, it first tries to insert the *last* key of the shorter file into the latter part of the longer file. By so doing, n and m can always be used to indicate the number of items remaining in A and B, respectively. (If the output file C overlaps file A to save space, it must extend from the end of A rather than the beginning as in Fig. 2.14.) BINARY MERGE will always do the binary searches on files of size $2^t - 1$ for some t, rather than exactly $\lfloor n/m \rfloor$. Thus, since the keys in the files to be merged will usually be distinct, the search will require exactly t comparisons. After the binary search finds the proper position for, say, b_m among $2^t - 1$ keys in the latter part of A, not all of those $2^t - 1$ keys can be moved to the output file because b_{m-1} might also belong in that segment of the file. Thus if $a_j < b_m < a_{j+1}$, then $b_m, a_{j+1}, \ldots, a_n$ are moved to C.

The trick of BINARY MERGE is to think of the two files playing different roles; the keys in the shorter file are merged into their proper place among the keys in the longer file. As the merging process continues, the file that was originally longer may become shorter. Thus m and n are compared and the roles played by A and B are reversed whenever necessary.

Let BSEARCH (x, F, ℓ, r, j) be a binary search algorithm whose arguments are:

x	A key
F	A file (either A or B in the merge algorithm) with individual entries in the file denoted by a subscripted "f"
ℓ and r	Positive integers indicating the boundaries of the subfile of F to be searched for x
j	An integer variable set by BSEARCH such that $\ell - 1 \leqslant j \leqslant r$ and $f_j \leqslant x \leqslant f_{j+1}$. (If $\ell > r$, j is set to r.)

Let MOVE(F, q, r) be an algorithm to move the qth through rth keys in the file F to the file C.

Algorithm 2.16 BINARY MERGE

Input: A and B, files with keys in nondecreasing order; and n and m, the number of keys in each.

Output: C, a file containing all the keys from A and B in nondecreasing order.

1. $p \leftarrow n + m$ [p is the index of the next vacant position in C.]
2.   **while** $n > 0$ and $m > 0$ **do**
3.   **if** $n > m$ **then** BMRGE (A, n, B, m)
4.   **else** BMRGE (B, m, A, n)
     end
5.   **if** $n = 0$ **then** MOVE $(B, 1, m)$
6.   **else** MOVE $(A, 1, n)$
7.   **stop**

* * * * * * * * * *

BMRGE (F, v, G, w)

Comment: F and G are files with v and w keys, respectively. BMRGE assumes $v \geqslant w$.

1. $k \leftarrow 2^{\lfloor \log (v/w) \rfloor}$
2.   **if** $g_w \leqslant f_{v-k+1}$
3.   **then do** MOVE $(F, v - k + 1, v)$
4. $v \leftarrow v - k$; $p \leftarrow p - k$
         end
5.   **else do** BSEARCH $(g_w, F, v - k + 2, v, j)$
6.   MOVE $(F, j + 1, v)$
7. $p \leftarrow p - v + j$
8. $c_p \leftarrow g_w$
9. $v \leftarrow j$; $w \leftarrow w - 1$; $p \leftarrow p - 1$
         end

Example 2.1   Assume that $n = 100$ and $m = 3$. Then $k = 2^{\lfloor \log 33.3 \rfloor} = 32$ and $n - k + 1 = 69$, so b_3 is compared to a_{69}. Suppose $b_3 > a_{69}$.

Search here for b_3.

A a_1 ... a_{69} a_{74} ... a_{100}

Suppose BSEARCH returns $j=74$.
Move these and b_3 to C.

Now $n = 74$, $m = 2$, $k = 2^{\lfloor \log 37 \rfloor} = 32$, and $n - k + 1 = 43$. Suppose $b_2 \leqslant a_{43}$.

A a_1 ... a_{43} --- a_{74}

Move these to C.

Now $n = 42$, $m = 2$, $k = 2^{\lfloor \log 21 \rfloor} = 16$, $n - k + 1 = 27$, and b_2 will be compared to a_{27} next.

WORST-CASE ANALYSIS OF BINARY MERGE

Let BM(m, n) be the maximum number of comparisons done by the BINARY MERGE algorithm when given files of sizes m and n. Clearly BM$(m, n) = $ BM(n, m) so without loss of generality we may assume $n \geqslant m$. We will establish a recurrence relation and boundary condition for BM in Theorem 2.12 and a closed-form formula for BM in Theorem 2.13. Theorem 2.12 is proved by double induction on m and n. That is, assuming the formulas hold for all pairs (m', n') less than (m, n) in lexicographic order, we will verify them for (m, n). In lexicographic order, $(m_1, n_1) < (m_2, n_2)$ if and only if $m_1 < m_2$ or $m_1 = m_2$ and $n_1 < n_2$.

Theorem 2.12 The recurrence relation for BM is

$$BM(m, 2n + \epsilon) = BM(m, n) + m \qquad \text{for } m \leqslant n, \ \epsilon = 0 \text{ or } 1 \qquad (2.1)$$

with boundary condition

$$BM(m, n) = m + n - 1 \qquad \text{for } m \leqslant n < 2m. \qquad (2.2)$$

Proof. The basis for the induction is to verify Eqs. (2.1) and (2.2) for the pair $(1, 1)$. This is trivial since an examination of the algorithm shows that $BM(1, 1) = 1$ and $BM(1, 2) = BM(1, 3) = 2$. The proof below is not valid for $m = 1$, so the verification of Eqs. (2.1) and (2.2) when $m = 1$ is left as an exercise (Exercise 2.21).

Now, observe that the first comparison performed in the algorithm (line 2 of BMRGE) reduces the problem to either merging m keys with $n - k$ keys, or doing a binary search (doing $\lfloor \log n/m \rfloor$ comparisons), and then merging $m - 1$ keys with j keys where $n - k + 1 \leqslant j \leqslant n$. In the latter case the maximal number of keys remains if $j = n$. Let $t = \lfloor \log n/m \rfloor$ and note that $k = 2^t$. Thus

$$BM(m, n) = 1 + \max \{BM(m, n - 2^t), t + BM(m - 1, n)\}, \qquad (2.3)$$

where $2^t m \leqslant n < 2^{t+1} m$. Substituting $2n + \epsilon$ for n, where $\epsilon = 0$ or 1, in (2.3) yields

$$BM(m, 2n + \epsilon) = 1 + \max \{BM(m, 2(n - 2^t) + \epsilon), t + 1 + BM(m - 1, 2n + \epsilon)\}, \qquad (2.4)$$

where $2^t m \leqslant n < 2^{t+1} m$. Equation (2.4) is used extensively in the proof. Now suppose that $1 < m = n$. Using Eq. (2.3) with $t = 0$ and the induction hypothesis for Eq. (2.2), $BM(m, m) = 1 + \max \{BM(m, m - 1), BM(m - 1, m)\} = 1 + BM(m - 1, m) = 1 + 2m - 2 = 2m - 1$, thus establishing Eq. (2.2) in this case. Using Eq. (2.4) with $t = 0$, $BM(m, 2m + \epsilon) = 1 + \max \{BM(m, 2m - 2 + \epsilon), 1 + BM(m - 1, 2m + \epsilon)\}$. Using the induction hypothesis for Eq. (2.2) gives $BM(m, 2m - 2 + \epsilon) = 3m - 3 + \epsilon = BM(m, m) + m - 2 + \epsilon$. By the induction assumption for Eq. (2.1) $1 + BM(m - 1, 2m + \epsilon) = 1 + BM(m - 1, m) + m - 1 = BM(m, m) + m - 1$. So $BM(m, 2m + \epsilon) = 1 + \max \{BM(m, m) + m - 2 + \epsilon, BM(m, m) + m - 1)\} = BM(m, m) + m$, thus establishing Eq. (2.1) for $m = n$. Now suppose $m < n$. For $m < n < 2m$, $BM(m, n) = 1 + \max \{BM(m, n - 1), BM(m - 1, n)\}$. By the induction assumption, the first term in the brackets is $m + n - 2$. To show that $BM(m, n) = m + n - 1$, we must have the second term in the brackets bounded by $m + n - 2$. If $n < 2(m - 1)$, then we can use Eq. (2.2) to get $BM(m - 1, n) = m + n - 2$. If $n = 2m - 2$ or $n = 2m - 1$, we write n as $2p + \epsilon$ where $p = m - 1$ and $\epsilon = 0$ or 1 and use Eqs. (2.1) and then (2.2) to get $BM(m - 1, 2p + \epsilon) = BM(m - 1, m - 1) + m - 1 = 2(m - 1) - 1 + m - 1 = m + n - \epsilon - 2 \leqslant m + n - 2$. Finally, we establish Eq. (2.1) for $m < n$. By Eq. (2.4), $BM(m, 2n + \epsilon) = 1 + \max \{BM(m, 2(n - 2^t) + \epsilon), t + 1 + BM(m - 1, 2n + \epsilon)\}$, where

$2^t m \leqslant n < 2^{t+1} m$. We can use the induction assumption for Eq. (2.1) on the terms in the brackets if $n - 2^t \geqslant m$. It is left as an exercise (Exercise 2.21) to verify that this inequality holds for $m > 1$. Therefore, $\mathrm{BM}(m, 2n + \epsilon) = 1 + \max \{\mathrm{BM}(m, n - 2^t) + m, t + 1 + \mathrm{BM}(m - 1, n) + m - 1\} = 1 + \max \{\mathrm{BM}(m, n - 2^t), t + \mathrm{BM}(m - 1, n)\} + m = \mathrm{BM}(m, n) + m$ by Eq. (2.3). This completes the induction proof of Eqs. (2.1) and (2.2). $\qquad\square$

Theorem 2.13 For $m \leqslant n$ and t such that $2^t m \leqslant n < 2^{t+1} m$,

$$\mathrm{BM}(m, n) = m + \left\lfloor \frac{n}{2^t} \right\rfloor - 1 + tm. \tag{2.5}$$

Thus since $t = \lfloor \log n/m \rfloor$, $\mathrm{BM}(m, n) \approx m(\log (n/m) + 2)$.

Proof. The proof of Eq. (2.5) is by induction on n. For $m \leqslant n < 2m$, (2.5) reduces to (2.2) because $t = 0$. Suppose $n \geqslant 2m$. Then for some $p \geqslant m$ and $\epsilon = 0$ or 1, $n = 2p + \epsilon$. Using Eq. (2.1) and the induction hypothesis,

$$\begin{aligned} \mathrm{BM}(m, n) &= \mathrm{BM}(m, 2p + \epsilon) \\ &= \mathrm{BM}(m, p) + m \\ &= m + \left\lfloor \frac{p}{2^t} \right\rfloor - 1 + t'm + m, \end{aligned}$$

where $t' = \lfloor \log p/m \rfloor = t - 1$. Since

$$\left\lfloor \frac{n}{2^t} \right\rfloor = \left\lfloor \frac{p}{2^{t-1}} \right\rfloor + \left\lfloor \frac{\epsilon}{2^t} \right\rfloor = \left\lfloor \frac{p}{2^{t-1}} \right\rfloor = \left\lfloor \frac{p}{2^t} \right\rfloor,$$

we have $\mathrm{BM}(m, n) = m + \lfloor n/2^t \rfloor - 1 + tm$. $\qquad\square$

Thus, when n is much larger than m, the number of comparisons done by BINARY MERGE is much lower than the $n + m - 1$ done by the straightforward merge algorithm (Algorithm 2.15) in the worst case.

LOWER BOUND FOR MERGING SORTED FILES

To derive a lower bound on the number of comparisons needed in the worst case to merge files of sizes n and m, we use decision trees similar to the decision trees used to obtain lower bounds for sorting and for searching ordered lists. Each node in the decision tree for a merge algorithm is labeled by a pair (i, j) to indicate the comparison of a_i with b_j; each leaf is labeled with a possible outcome. To compute the number of leaves, observe that each outcome corresponds to a choice of m positions (for the keys in B) from the $n + m$ positions in C. Thus the number of leaves is $\binom{n+m}{m}$ and the number of comparisons done in the worst case is at least $\lceil \log \binom{n+m}{m} \rceil$. If $m = 1$, the lower bound is $\lfloor \log n \rfloor + 1 = \mathrm{BM}(1, n)$. If $n = m$, the lower bound is approximately $2n - \frac{1}{2} \log n$.[*] For this case we have already established a better lower bound of $2n - 1$

---

[*]   This result is obtained by using Stirling's approximation for the factorial function.

(which is achieved by the binary merge algorithm), so we can conclude that $\lceil \log \binom{n+m}{m} \rceil$ comparisons are not sufficient for merging in all cases. However, it has been shown that

$$\text{BM}(m, n) < \lceil \log \binom{m+n}{m} \rceil + \min \{m, n\}.$$

2.9   *External Sorting*

We have seen that bucket-sorting algorithms may do few or no comparisons of keys. It is therefore appropriate to separate them from the class of algorithms studied at the beginning of this chapter. There is a class of algorithms that does sort by comparison of keys but that is also distinguished from those studied earlier because the number of comparisons done is not the most appropriate measure of the efficiency of the algorithms. These are the *external sorts*. Here we assume that the number of keys is so large that they cannot all fit in the computer's high-speed memory at one time; some must be stored on a slower external storage device. The time required to transfer data back and forth between the high-speed memory and the external storage device usually outweighs the time required to perform comparisons on data in the high-speed memory.

Throughout this section we will refer to the external storage devices as tapes. This is done solely to simplify the language. The methods presented can be used with disks and drums with minor changes, if any. With such storage devices it is most efficient to process the keys sequentially because tape rewinds, disk seeks, and other operations needed to locate a particular datum are very expensive in terms of time. In present practice, the data for an external sort would usually be on a disk rather than on tapes, and it would probably be scattered in blocks in a number of different locations on the disk. A significant amount of time is required for the disk seeks (i.e., to locate the appropriate block). Hence the emphasis in devising good algorithms is on decreasing the number of times the data are accessed.

POLYPHASE MERGE SORTING

In the previous section we studied algorithms to merge sorted files. The straightforward merge algorithm made one pass over the data, processing the keys of each of the two files to be merged in sequential order. The algorithm is easily adapted to merge files that are on two tapes. The external sorting methods that we will study in this section rely heavily on the merge algorithm. Because merging must be done repeatedly, these algorithms are called polyphase merge sorting algorithms.

We define a *run* as a sequence of keys in nondecreasing order. The general scheme of the external sorts has two phases:

(I)   Run construction and distribution: arrange the keys in runs on two (or more) tapes; and

(II) Polyphase merge: repeatedly merge the runs until there is only one.

Let m be the number of keys that can fit in the high-speed memory at one time along with whatever programs are necessary. (To some extent, m will depend on the algorithm being used. In describing sorting algorithms throughout this chapter, we have usually ignored the fact that each key is associated with a perhaps very large record. This point is relevant here since if the records are large, m doesn't vary very much from one algorithm to another.) As usual, let n be the total number of keys.

The first, and simplest, external sorting algorithm uses four tapes: T_0, T_1, T_2, and T_3. The keys to be sorted are initially on T_0. The following outline describes the algorithm.

Algorithm 2.17 EXTERNAL SORT WITH FOUR TAPES (outline)
[Phase I: Run construction and distribution]

while there are more records on T_0 **do**
    1. Read in m records (perhaps less the last time).
    2. Sort them using an internal sorting algorithm.
    3. If the previous run was put on T_2, put this one on T_3; else on T_2.
end

Rewind the tapes.
[Phase II: Merging the runs]

$i_1 \leftarrow 2; i_2 \leftarrow 3$ [indexes of input tapes]
$j_1 \leftarrow 0; j_2 \leftarrow 1$ [indexes of output tapes]

while there is more than one run **do**
    1. Merge the first run on T_{i_1} with the first run on T_{i_2} and put the resulting run on T_{j_1}.
    2. Merge the next run from T_{i_1} and T_{i_2} and put the result on T_{j_2}.
    3. Repeat steps 1 and 2 (putting the output alternately on T_{j_1} and T_{j_2}) until the end of data on T_{i_1} and T_{i_2} is reached.
    4. Rewind the tapes. Add 2 (modulo 4) to i_1, i_2, j_1, and j_2 [to reverse the roles of input and output tapes].
end

At various times there may be one more run on T_{i_1} than on T_{i_2}. The odd run would simply be copied onto the appropriate output tape. Figure 2.15 illustrates the algorithm.

To analyze this and similar algorithms, we will count several operations: comparisons of keys, tape rewinds, and complete passes over the data. In one execution of the loop in the merge phase (lines 1 through 4), each record is transferred from an input tape to the high-speed memory where some operations (comparisons of keys) are performed, and then is transferred to an output tape. Then the tapes are rewound (simultaneously). Thus the number of rewinds and passes over the data in the merge phase is the number of passes through the loop. Let k be the number of keys in each run formed during the run construction phase. (To simplify the analysis we will assume all of the initial runs have the same length; in fact, the last one may be smaller.) Then the total number of runs on the tapes when the merge phase begins is $r = n/k$.

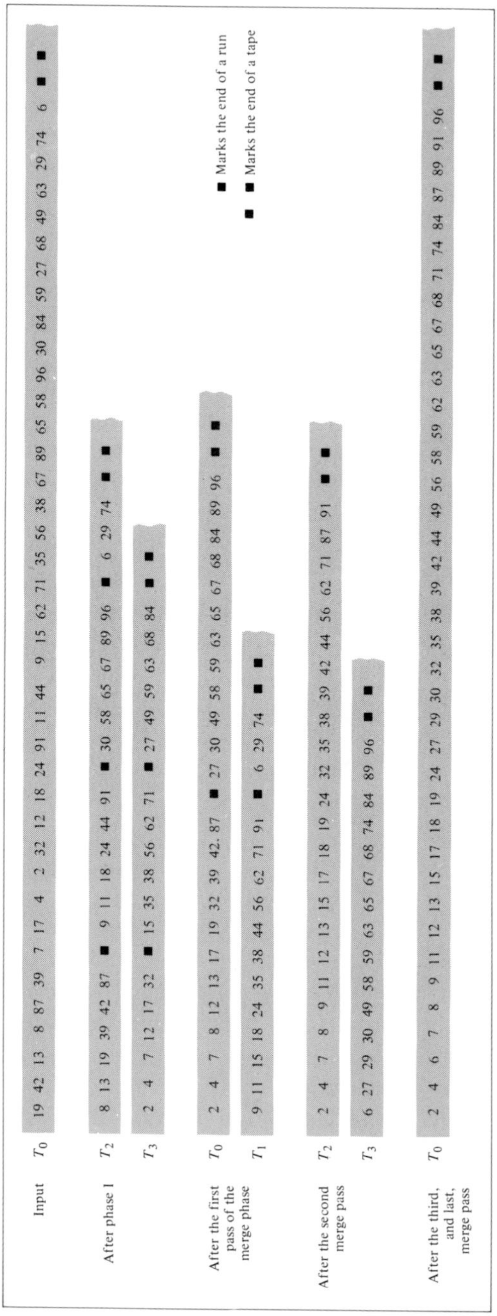

Figure 2.15
Polyphase merge with four tapes, $m = 6$.

After the first merge pass there is a total of $\lceil r/2 \rceil$ runs on T_0 and T_1. After the next merge pass there are $\lceil r/4 \rceil$ runs, and after the ith merge pass, there are $\lceil r/2^i \rceil$. The algorithm terminates when there is only one run so the number of merge passes is $\lceil \log r \rceil$. Since the run construction phase does just one pass over the data followed by one rewind (of three tapes simultaneously), the total number of passes is $\lceil \log r \rceil + 1$. Recall $r = n/k$ and in Algorithm 2.17 k of course is m, the number of keys that can be sorted in the high-speed memory. If we are working with a specific computer, m cannot be changed, but we see that the number of passes would be smaller if the merge phase started with fewer, larger runs, that is, if $k > m$. This observation motivates an important improvement in external sorting that will be described later.

The number of comparisons done depends on the sorting algorithm used in the run construction phase. Suppose an $\Theta(m \log m)$ sort is used. Then the number of comparisons done in the sorting phase is $\Theta[(n/m) \cdot m \log m]$, or $\Theta(n \log m)$. In the first merge pass $r/2$ pairs of files, each of size k, are merged, requiring at most $(r/2)(2k - 1)$ comparisons. The ith merge pass requires at most $(r/2^i)(2^i k - 1) = n - (r/2^i)$ comparisons, so the total number of comparisons done in the merge phase is

$$\sum_{i=1}^{\lceil \log r \rceil} \left(n - \frac{r}{2^i} \right) = n \lceil \log r \rceil - r + \beta,$$

where $\frac{1}{2} < \beta \leqslant 1$. Observe that the number of comparisons done in the merge phase would also be reduced if r were reduced, i.e., if k could be increased. For Algorithm 2.17, however, with $k = m$, the total number of comparisons done is $\Theta[n \log m + n \log (n/m)]$, which is $\Theta(n \log n)$. The total number of passes over the n keys is $\lceil \log(n/m) \rceil + 1$.

POLYPHASE MERGING WITH THREE TAPES

We have assumed a specific set of resources here: a high-speed memory that can hold m records and four tape drives. An attempt to improve or modify the external sort can proceed in any of several directions. We may try to find a more efficient algorithm using the same resources, or we may assume that less, or more, tape drives are available. In fact many systems do not have four tape drives, so we will now consider the problem of doing the merge phase with three tapes.

We assume that all records were initially on T_0 and that in the first phase r runs of size k were constructed and written on the other two tapes, T_1 and T_2. Each merge pass must begin with the runs on two "input" tapes. We can merge runs from the input tapes as before, but since there are only three tapes, all of the resulting runs will go on the one output tape. Then before beginning the next merge, half of the runs are copied onto one of the input tapes. Thus Phase II of the algorithm would be rewritten as follows:

Phase II' Merging the runs using three tapes

$i_1 \leftarrow 1; i_2 \leftarrow 2$ [indexes of "input" tapes]

$j \leftarrow 0$ [index of "output" tape]

while there is more than one run **do**

   1'   Merge runs from T_{i_1} and T_{i_2} and put the resulting runs on T_j.

   2'   Rewind all three tapes. Copy half the runs from T_j onto T_{i_2}. Rewind T_{i_2}.

   3' $i_1 \leftarrow i_1 + 1$; $i_2 \leftarrow i_2 + 1$; $j \leftarrow j + 1$ (all modulo 3)

end

The number of comparisons done when three tapes are used is the same as the number of comparisons done with four tapes.

We define the number of passes over the data to be the average of the numbers of times keys are moved from one tape to another, i.e., the total number of keys moved divided by n. (For the four-tape sort every key was moved the same number of times so we did not need to state this definition earlier.) The number of passes in the three-tape merge will be larger than for the four-tape merge because of the extra partial passes for copying. There is one pass for the run construction phase; then $1.5 \lceil \log r \rceil - \frac{1}{2}$ passes in the merge phase. (The $-\frac{1}{2}$ reflects the fact that no keys must be copied after the final merge pass. That is, in practice the algorithm would terminate in line 2'.) Thus, using three tapes as described, the number of passes is $\frac{1}{2} + 1.5 \lceil \log r \rceil$.

Copying keys from one tape to another should seem like a very nonproductive effort. Can it be avoided if we are restricted to using only three tapes? Clearly if, in the run construction phase, the runs are distributed equally on the two output tapes, then after the first merge pass all of the keys will be on one tape and some must be copied. Suppose the runs are not distributed equally. For example, suppose eight runs are on tape T_1 and five on tape T_2, as in Fig. 2.16(a), where runs are shown as rectangles labeled with their size (number of keys). After merging five pairs of runs onto T_0, T_2 is empty and T_1 still contains three runs. Now, T_2 and T_0 may be rewound and three pairs of runs from T_1 and T_0 may be merged and placed on T_2. (See Figs. 2.16(b) and (c). Note that runs of different sizes are being merged.) After rewinding T_1 and T_2, two pairs of runs from T_2 and T_0 are merged and placed on T_1. After three more merge passes, the details of which may be worked out by the reader, there will be one run of size $13k$ on T_2 and the sort will be completed.

How many passes over the data were made in this example? To get from the situation in Fig. 2.16(a) to that in Fig. 2.16(b), only 10/13 of the keys are transferred from T_1 and T_2 to T_0; three runs remain untouched on T_1. To get to the situation in Fig. 2.16(c) 9/13 of the keys are moved. In the remaining steps not shown 10/13, then 8/13, and finally, on the last pass, 13/13 of the keys are moved. The total number of passes during the merge phase is 50/13, or approximately 3.8. If the 13 runs had been arranged with six on one tape and seven on another, and the Phase II' method with copying were used, it would do four complete passes over the data while merging runs and 22/13 copying passes for a total of 74/13, or approximately 5.7. At

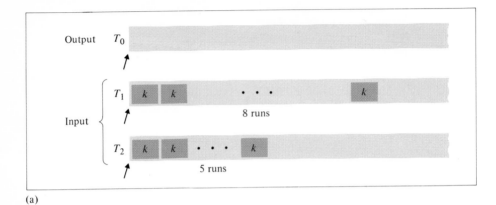

(a)

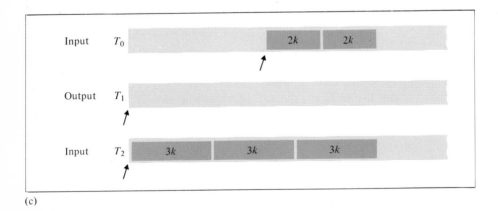

(b)

(c)

Figure 2.16
Merging with three tapes, but no copying. (Arrows indicate the location of the tape head.)

Table 2.2

Data on the example in Fig. 2.16. The total number of runs at the beginning of the merge phase is $13 = F_7$.

Runs on the two input tapes		Runs merged	
Number	Size (as a multiple of k)	Number of pairs	Total number of size k runs transferred from one tape to another
8, 5	1, 1	5	10
5, 3	2, 1	3	9
3, 2	3, 2	2	10
2, 1	5, 3	1	8
1, 1	8, 5	1	13
1	13		

no point in the example in Fig. 2.16 did we get "stuck" and have to copy keys; that is, at no point were all the keys on one tape before they were completely sorted. Were we just lucky this time? Or can we arrange the runs constructed in the first phase of the algorithm so that copying can always be avoided? Clearly, copying can be avoided by putting one run on T_2 and all the other $r-1$ runs on T_1, but this arrangement would require rewinding two tapes each time a pair of runs is merged. The total number of rewinds (of two or three tapes simultaneously) would be r. As an exercise, the reader may compute the total number of passes over the keys (Exercise 2.24). It is not $O(\log r)$. So, it is not difficult to avoid copying keys. Our goal, however, is to eliminate copying *and* to keep the number of passes over the keys relatively small.

Look at Table 2.2, which provides a summary of the example illustrated in Fig. 2.16. The table lists the number and size of the runs on the tapes after each rewind. The columns of numbers in the table are all sequences of Fibonacci numbers.[*] In general, if the number of runs constructed in the first phase of the sort is a Fibonacci number F_s, for some s, then distributing the runs so that F_{s-1} runs are on one tape and F_{s-2} are on another will ensure that there will always be runs on two tapes ready for merging without copying until the keys are sorted. Assuming that the run construction phase has put F_{s-2} runs on T_1 and F_{s-1} on T_2, lines 1' and 2' of the Phase II' algorithm for merging using three tapes would be changed to the following:

1″ Merge runs from T_{i_1} and T_{i_2}, putting the resulting runs on T_j, until T_{i_1} is empty.

2″ Rewind T_{i_1} and T_j.

---

[*] The Fibonacci numbers are a sequence of integers defined by the recurrence relation $F_i = F_{i-1} + F_{i-2}$ for $i > 1$, with boundary conditions $F_0 = 0$ and $F_1 = 1$.

Table 2.3

Runs on the input tapes				Merged runs	
Number$_1$	Size$_1$ (as a multiple of k)	Number$_2$	Size$_2$ (as a multiple of k)	Number of pairs	Number of keys transferred from one tape to another = Number of pairs × (Size$_1$ + Size$_2$), as a multiple of k
F_{s-1}	$1 = F_2$	F_{s-2}	$1 = F_1$	F_{s-2}	$F_{s-2}(F_2 + F_1) = F_{s-2}F_3$
F_{s-2}	$2 = F_3$	F_{s-3}	$1 = F_2$	F_{s-3}	$F_{s-3}(F_3 + F_2) = F_{s-3}F_4$
F_{s-3}	$3 = F_4$	F_{s-4}	$2 = F_3$	F_{s-4}	$F_{s-4}(F_4 + F_3) = F_{s-4}F_5$
$\cdots$		$\cdots$		$\cdots$	$\cdots$
F_{s-i}	F_{i+1}	$F_{s-(i+1)}$	F_i	$F_{s-(i+1)}$	$F_{s-(i+1)}(F_{i+1} + F_i) = F_{s-(i+1)}F_{i+2}$
$\cdots$		$\cdots$		$\cdots$	$\cdots$
$F_2 = 1$	F_{s-1}	$F_1 = 1$	F_{s-2}	F_1	$F_1(F_{s-1} + F_{s-2}) = F_1 F_s$
$F_1 = 1$	F_s	-------	-------		

With these changes, we will refer to the algorithm as Phase II″: merging with three tapes using a Fibonacci distribution. If initially there are F_s runs, the Phase II″ loop will be executed $s - 1$ times. To aid in computing the number of passes – i.e., the average number of times each key is transferred from one tape to another – we use Table 2.3, a generalized version of Table 2.2.

The number of passes is the sum of the righthand column divided by n (i.e., $F_s k$). Thus we must compute

$$\frac{\sum_{i=1}^{s-2} F_i F_{s-i+1} k}{F_s k} = \frac{\sum_{i=1}^{s-2} F_i F_{s-i+1}}{F_s} . \tag{2.6}$$

Lemma 2.14

$$\sum_{i=1}^{s-2} F_i F_{s-i+1} = \frac{s-5}{5} F_{s+1} + \frac{2s+2}{5} F_s, \qquad \text{for } s \geqslant 2.$$

Proof. The proof is by induction on s. In the induction step we will need the assumption that $s \geqslant 4$, so for the basis of the induction the equation in the lemma must be checked for $s = 2$ and $s = 3$. The reader may easily verify that for $s = 2$ both sides have value 0, and for $s = 3$ both sides have value 2.

Now, assuming $s \geqslant 4$, and using the recurrence relation for the Fibonacci numbers and the induction assumption, we have

$$\sum_{i=1}^{s-2} F_i F_{s-i+1} = \sum_{i=1}^{s-2} F_i (F_{s-i} + F_{s-i-1})$$

$$= \sum_{i=1}^{s-2} F_i F_{s-i} + \sum_{i=1}^{s-2} F_i F_{s-i-1}$$

$$= \sum_{i=1}^{s-2} F_i F_{(s-1)-i+1} + \sum_{i=1}^{s-2} F_i F_{(s-2)-i+1}$$

$$= \left(\sum_{i=1}^{s-3} F_i F_{(s-1)-i+1} + F_{s-2} \right) + \left(\sum_{i=1}^{s-4} F_i F_{(s-2)-i+1} + F_{s-3} + F_{s-2} \right)$$

$$= \left(\frac{(s-1)-5}{5} F_s + \frac{2(s-1)+2}{5} F_{s-1} + F_{s-2} \right)$$

$$\quad + \left(\frac{(s-2)-5}{5} F_{s-1} + \frac{2(s-2)+2}{5} F_{s-2} + F_{s-3} + F_{s-2} \right)$$

$$= \frac{s-6}{5} F_s + \frac{3s-2}{5} F_{s-1} + \frac{2s+3}{5} F_{s-2}$$

$$= \frac{3s-3}{5} F_s + \frac{s-5}{5} F_{s-1} = \frac{s-5}{5} F_{s+1} + \frac{2s+2}{5} F_s. \qquad \square$$

(The formula in Lemma 2.14 was not pulled out of a hat. The reader interested in its derivation may consult the notes and references at the end of this chapter.) Using Eq. (2.6) we see that the number of passes is

$$\frac{s-5}{5} \cdot \frac{F_{s+1}}{F_s} + \frac{2s+2}{5}.$$

We want to relate this expression to the number of runs, $r = F_s$, so that the number of passes done using the Fibonacci distribution can be compared to the number of passes done by the other methods discussed. We will use the following lemma, the proof of which is easy and is left as an exercise.

Lemma 2.15 Let F_j be the jth Fibonacci number. Then for $j \geqslant 5$,

$$\frac{F_{j+1}}{F_j} \leqslant \frac{13}{8}.$$

We also use the fact that

$$F_j \approx \frac{[\frac{1}{2}(1+\sqrt{5})]^j}{\sqrt{5}}. \tag{2.7}$$

(See the references. In fact, F_j is $[\frac{1}{2}(1+\sqrt{5})]^j/\sqrt{5}$ rounded to the nearest integer.)

Theorem 2.16 The number of passes done in Phase II″ (merging with three tapes using a Fibonacci distribution) is approximately $1.04 \log r$, where r is the number of runs produced by the sort phase.

Proof. Using Eq. (2.7), we have

$$\sqrt{5}F_j \approx [\frac{1}{2}(1+\sqrt{5})]^j \quad \text{and} \quad \log(\sqrt{5}F_j) = j[\log(1+\sqrt{5})-1].$$

So

$$j \approx \frac{\log\sqrt{5} + \log F_j}{\log(1+\sqrt{5})-1} \approx 1.67 + 1.43 \log F_j.$$

Therefore the number of passes is approximately

$$\frac{1.67 + 1.43 \log F_s - 5}{5} \cdot \frac{13}{8} + \frac{2(1.67 + 1.43 \log F_s) + 2}{5} \approx 1.04 \log F_s$$

$$= 1.04 \log r. \qquad \square$$

The result in Theorem 2.16 compares quite favorably with the number of passes done by Phase II′, where runs had to be copied from one tape to another. In fact, using the Fibonacci distribution is almost as good as using four tapes. The Fibonacci distribution is also very useful if *more* tapes are available; it can be generalized to produce very fast algorithms. With six tapes, for example, the number of passes would be roughly $0.55 \log r + 0.86$.

ARRANGING THE RUNS FOR MERGING

We have seen that if the number of sorted runs is a Fibonacci number, say F_s for some s, the number of merge passes will be relatively low if the runs are distributed so that F_{s-1} are on one tape and F_{s-2} on another. But the total number of runs may not be known in advance, and it may not be a Fibonacci number. Also, as we shall see later, the size of each run, and hence the total number, may depend on the particular sequence of keys to be sorted. Thus our fast merge phase algorithm is not of much use without an efficient way of arranging the runs produced by the run construction phase to ensure that their arrangement on the tapes approximates the Fibonacci arrangement. This is accomplished by writing the runs on the tapes so that first there is one run on one tape and none on the other, then two on one tape and one on the other, then three and two, then five and three, and so on, so that after each round (except perhaps the last) the numbers of runs on the tapes are two consecutive Fibonacci numbers. The writing out of runs ends when there are no more keys. At that time the number of runs on one tape is between F_{s-2} and F_{s-1}, and the number of runs on the other is between F_{s-3} and F_{s-2}. The merge phase can then act as though there are enough "dummy" runs on each tape to give F_{s-1} and F_{s-2} runs, respectively. By keeping track of the number of dummy runs and treating them appropriately, the merge phase can be made to mimic a merge with F_s runs.

To write the algorithm for arranging the runs, we must determine how many runs are to be added to each tape during each round of writing. Let R_i^j be the number of runs on T_i after j rounds. Then

$$R_1^1 = 1 = F_1, \qquad R_2^1 = 0 = F_0,$$

and

$$R_1^j = F_j, \qquad R_2^j = F_{j-1}, \text{ for } j > 1.$$

Let D_i^j be the number of runs to be added to T_i during round j. $D_1^j = R_1^j - R_1^{j-1} = F_j - F_{j-1} = F_{j-2} = R_2^{j-1}$, and $D_2^j = R_2^j - R_2^{j-1} = R_1^{j-1} - R_2^{j-1}$. Thus D_i^j, for $i = 1, 2$, can easily be computed from R_i^{j-1}, $i = 1, 2$. Also, of course, $R_1^j = R_1^{j-1} + R_2^{j-1}$ and $R_2^j = R_1^{j-1}$. In the algorithm the superscripts are omitted; the values of D_1, D_2, R_1, and R_2 are updated after each round of writing. A straightforward algorithm would write out D_1 runs on T_1, then D_2 runs on T_2, then update the values of the D's and R's and begin the next round of run construction and writing. It would halt when there are no more runs to be written on the tapes. If the runs are written on the tapes in this manner, it is possible that when the algorithm terminates all of the dummy runs will be on T_2. "Merging" a run with a dummy means just copying the run from an input tape to the output tape. "Merging" two dummy runs requires only that the number of dummies on each input tape be decremented and the count of dummies on the output tape be incremented. Hence, when possible, we would like the number of dummies on each tape to be equal. At the beginning of each round of writing, $D_1^j = F_{j-2}$ and $D_2^j = F_{j-3}$. Since D_1^j is much larger than D_2^j, the scheme is to write runs on T_1 until $D_1^j = F_{j-3}$, and then alternate between the two tapes. Thus we have the following algorithm.

Algorithm 2.18 FIBONACCI RUN DISTRIBUTION

1. $R_1 \leftarrow R_2 \leftarrow 0; D_1 \leftarrow 1; D_2 \leftarrow 0$
       while $1 = 1$ **do**
2.   **while** $D_1 \neq D_2$ **do**
               if END-OF-INPUT **then return**
               Write a run on T_1
               $D_1 \leftarrow D_1 - 1$
           end
3.   **while** $D_2 > 0$ **do**
               for $i \leftarrow 1$ **to** 2 **do**
                   if END-OF-INPUT **then return**
                   Write a run on T_i
                   $D_i \leftarrow D_i - 1$
               end
           end
4. $D_1 \leftarrow R_2; D_2 \leftarrow R_1 - R_2;$
           $RR \leftarrow R_1; R_1 \leftarrow R_1 + R_2; R_2 \leftarrow RR$
       end

As indicated in the discussion above, a count of the number of dummy runs on each tape is maintained during the merge phase.

RUN CONSTRUCTION

Recall that our general external sorting scheme consists of the following two phases: (I) run construction and distribution; and (II) polyphase merging. For each of the several variations we have considered for the merge phase, the analysis has shown that the number of passes over the keys would decrease if the size of the runs constructed in the first phase could be increased. Our approach to the run construction was very simple and straightforward. Assuming m is the number of keys that fit in the high-speed memory at one time, we read in m keys, sort them using a fast internal sort, and output a run of size m. All of the runs, except perhaps the last, will have exactly m keys. Now we will consider an alternative method of constructing runs which is expected to produce longer ones. This method is called replacement selection. The main idea behind it is that as soon as one record has been written on one of the output tapes, another may be read in from the input tape. If its key is larger than the key just written, the new record may be made part of the current run. If its key is smaller, it will be held for the next run. Thus records may be added to the current run until all of memory is filled with records whose keys are smaller than the last key put in the run. At that point the current run is ended and the smallest key in memory is the first key of the next run. While a run is being constructed, we may think of the keys in memory as being divided into two classes: the active ones, those that are at least as large as the last key written out and that will be part of the current run, and the inactive keys, those that are smaller than the last key written and that will be part of the next run. Figure 2.17 illustrates the run construction. Inactive keys are shown in parentheses.

Input tape

Front of tape

11, 91, 24, 18, 12, 32, 2, 4, 17, 7, 39, 87, 8, 13, 42, 19

Memory (m = 6)

Output tape

Front of tape

Read in m records:

Input tape	Memory (m = 6)	Output tape
11, 91, 24, 18, 12, 32, 2, 4, 17, 7	39 87 8 13 42 19	8
11, 91, 24, 18, 12, 32, 2, 4, 17	39 87 (7) 13 42 19	13, 8
11, 91, 24, 18, 12, 32, 2, 4	39 87 (7) 13 42 19	17, 13, 8
11, 91, 24, 18, 12, 32, 2	39 87 (7) (4) 42 19	19, 17, 13, 8
11, 91, 24, 18, 12, 32	39 87 (7) (4) 42 (2)	39, 19, 17, 13, 8
11, 91, 24, 18, 12	(32) 87 (7) (4) 42 (2)	42, 39, 19, 17, 13, 8
11, 91, 24, 18	(32) 87 (7) (4) (12) (2)	87, 42, 39, 19, 17, 13, 8
11, 91, 24	(32)(18)(7)(4)(12)(2)	■ 87, 42, 39, 19, 17, 13, 8

Begin a new run.

Input tape	Memory (m = 6)	Output tape
11, 91, 24	32 18 7 4 12 2	2, ■ 87, 42, 39, 19, 17, 13, 8
11, 91	32 18 7 4 12 24	4, 2, ■ 87, 42, 39, 19, 17, 13, 8
11	32 18 7 91 12 24	7, 4, 2, ■ 87, 42, 39, 19, 17, 13, 8
	32 18 11 91 12 24	

Output remaining keys in order:

■ 91, 32, 24, 18, 12, 11, 7, 4, 2, ■ 87, 42, 39, 19, 17, 13, 8

Figure 2.17
Run construction using replacement selection. (■ marks the end of a run.)

How can we easily distinguish between active and inactive keys in the run construction algorithm? What methods and data structure should be used?

Constructing a run requires that we first find the smallest key in a set of keys and output it, then find the next smallest, and so on. In Section 1.5 we studied the tournament method for finding the smallest key in a file and we observed that to find the second key we merely had to examine (i.e., compare) those keys that "lost" to the smallest. (Actually in Section 1.5 we were looking for the largest and second largest keys, but the algorithm can obviously be adapted to the present situation.) To construct runs we must continually find the "next" smallest key and output it. The data structure and algorithm in Section 1.5 are not sufficient for this. They will be modified and extended here. We use a tree in which the leaves contain the records and keys as they are read in and the internal nodes contain pointers to the loser (rather than the winner, as in Section 1.5) of each comparison. The field names for these data are RECORD, KEY, and LOSER, respectively. The tournament on the first six keys in the example in Fig. 2.17 could be represented by the tree in Fig. 2.18(a).

Each time a record is written on a tape, a new one is read in and the structure must be updated. To describe the core of the algorithm, the updating of the loser pointers, we will temporarily ignore the problem of initializing the tree and the case when the new key must wait for the next run because it is too small to fit in the current one. Let W (for winner) point to the leaf containing the smallest key. The LOSER pointers are updated in the following routine, which also resets W to point to the leaf containing the next smallest key.

```
output RECORD(W)
RECORD(W) ← next record from the input tape
T ← PARENT(W)
while T ≠ Λ do
        if KEY(LOSER(T)) < KEY(W) then LOSER(T) ↔ W
        T ← PARENT(T)
end
```

What shall we do with the inactive records, in other words, those that are not part of the current run? A simplistic answer is to mark them somehow, skip over them during the updating procedure just described, and, finally, when all the records in memory are so marked, set up a new tournament tree and begin working on a new run. This solution is inefficient and certainly inelegant. Fortunately we can do better. Suppose we associate a run number (RN) with each record in memory. The number can easily be assigned when the record is read in; if its key is larger than the key just written out, its run number is the same as the latter's; otherwise, it is one higher. Each key may be considered as a pair (RN, KEY) where RN is the run number and KEY is the original key. When the LOSER pointers are updated, run numbers, as well as keys, are compared, and the pair (RN_1, KEY_1) is considered smaller than (RN_2, KEY_2) if $RN_1 < RN_2$ or if $RN_1 = RN_2$ and $KEY_1 < KEY_2$. Observe that this slight additional piece of information, along with small changes in the updating algorithm,

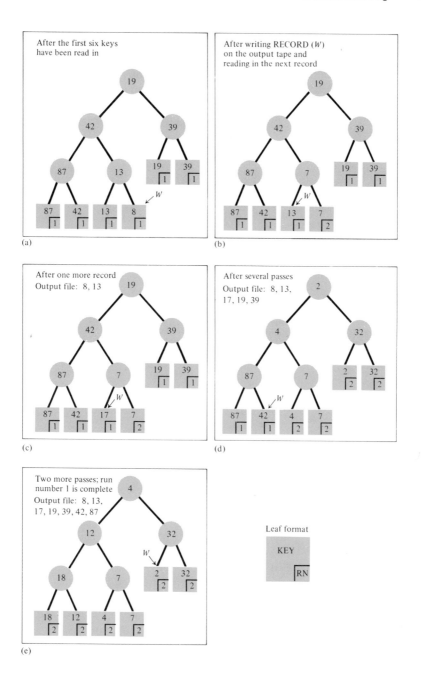

Figure 2.18
Trees for run construction. (The internal nodes contain pointers to the keys shown in the node.)

allows us to correctly manipulate keys from different runs in the same tree at one time. When the tree is filled with keys that must go in a new run, the LOSER pointers are already set and no initialization need be done to the data structure. The revised pointer updating algorithm that handles the case when the new record does not belong in the current run is presented below. Here we assume that there is an RN field in each leaf and R is the number of the run currently being constructed. W, as before, points to the next record to be written out.

[Output and input]
1. **if** RN(W) $\neq R$
2.   **then do** whatever is necessary for beginning a new run on the output tapes, including determining which tape the next run goes on.
3. $R \leftarrow$ RN(W)
            end
4. LAST $\leftarrow$ KEY(W)
5. **output** RECORD(W)
6. RECORD(W) $\leftarrow$ next record from the input tape
7. **if** KEY(W) $<$ LAST **then** RN(W) $\leftarrow R + 1$
                    [else RN(W) is still R as before]
[Update LOSER pointers]
8. $T \leftarrow$ PARENT(W)
9. **while** $T \neq \Lambda$ **do**
10.   **if** RN(LOSER(T)) $<$ RN(W) or
11.   RN(LOSER(T)) $=$ RN(W) and KEY(LOSER(T)) $<$ KEY(W)
12.   **then** LOSER(T) $\leftrightarrow W$
13. $T \leftarrow$ PARENT(T)
14. **end**

In Fig. 2.18 the tree is shown at several stages of the run construction for the example in Fig. 2.17. It is strongly suggested that the reader begin with the tree in Fig. 2.18(a), with $R = 1$, and follow the steps of the algorithm to see how the structure is modified to produce the remaining trees in Fig. 2.18.

Now how do we get the tree in Fig. 2.18(a) set up in the first place? The trick to the initialization is to start with a tree with "empty" nodes and assume the leaves contain keys from run number 0, a run that will not be written on the output tape. The LOSER pointers may be initialized in any way such that each of the leaves except one is pointed to by exactly one LOSER pointer. W is initialized to point to the one leaf that is not a loser. (In Fig. 2.18(a) W was initially set to the location of the first leaf, i.e., the one containing the key 19, and the LOSER pointers were assigned in order, left to right, as shown in Fig. 2.19.) Each time a new record is read in, exactly one LOSER pointer that had pointed to a leaf with RN = 0 will be modified to point to a leaf containing a record. Until the tree is full, W will point to leaves with RN = 0, thus indicating an "empty" leaf where the next record may be put.

Terminating the run construction algorithm smoothly requires the use of another "dummy" run; when there are no more records on the input tape, the run number of

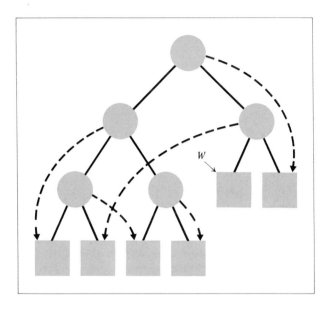

Figure 2.19
Initialization for the tree in Fig. 2.18(a). Loser pointers
are shown as dotted lines.

the leaf to be filled is incremented just as if the new key did not belong to the current
run. A counter is used to keep track of the number of real runs. Thus, whenever a
record is to be written on an output tape, its run number must be tested to see if it is
0 or if it is higher than the maximum run number (the value of which is not known in
advance, since it depends on the number and arrangement of the keys on the input
tape). With the initialization and termination details added to the pointer updating
algorithm, we have the complete replacement selection algorithm.

Algorithm 2.19 REPLACEMENT SELECTION
Comment: RUNS is a counter for the number of real runs.
$\qquad$ R is the current run number.
[Initialization]
RUNS $\leftarrow$ 0; $R \leftarrow$ 0; LAST $\leftarrow \infty$
Initialize LOSER pointers and W
Set RN to 0 for each leaf
[Output, input, and update loser pointers]
while RN(W) $\leqslant$ RUNS **do**
$\qquad$ **if** RN(W) $\neq R$
$\qquad\qquad$ **then do** whatever is necessary for beginning a new run on the output tapes,
$\qquad\qquad\qquad$ including determining which tape the next run goes on.
$\qquad\qquad\qquad$ $R \leftarrow$ RN(W)
$\qquad\qquad$ **end**

> **if** RN(W) $\neq$ 0 **then do** LAST $\leftarrow$ KEY(W)
>   **output** RECORD(W)
>   **end**
> **if** input is exhausted
>   **then** RN(W) $\leftarrow$ RUNS + 1 [dummy run]
>   **else do** RECORD(W) $\leftarrow$ next record from the input tape
>   **if** KEY(W) $<$ LAST
>   **then do** RN(W) $\leftarrow R$ + 1
>   RUNS $\leftarrow$ RN(W)
>   **end**
>   **end**
> Update LOSER pointers [lines 8–14 from the earlier version of the algorithm].
> **end**

Note that although Algorithm 2.18, the run distribution algorithm, was written independently of the run construction algorithm, the two must be properly interwoven. (The details are left to the reader.) There are also details to be handled for buffering the input and output; in practice, one does not read or write only one record at a time.

There are also, of course, details to be worked out for the representation of the tree. Paths in the tree are traversed only in an upward direction – i.e., from leaf to root – so from each node we need a way of finding the location of its parent. Since the tree is balanced we can use a storage scheme similar to the one used for heaps; that is, store the nodes sequentially so that the parent of the ith node is the $\lfloor i/2 \rfloor$ th node. Allocating space for the tree structure and determining the exact location of the parent of a node is complicated by the facts that the leaves, which contain the records, require much more space than the internal nodes, and the size of the records is probably not known when the program is written. A clever storage layout solves these problems: Each leaf is stored with one internal node so that there are, say, m_0 large nodes. (Since there are m_0 leaves and only $m_0 - 1$ internal nodes, a dummy internal node number 0 is used.) Thus if the tree nodes are numbered consecutively from the root (1) to the last leaf ($2m_0 - 1$), then for each i such that $0 \leqslant i < m_0$, node i, an internal node, is stored with node $m_0 + i$, a leaf. The location of the parent of the internal node and the location of the parent of the leaf may be computed when the structure is initially set up and stored with the data. Figure 2.20 illustrates a possible format for the double nodes in this structure. If this storage arrangement is used, W would be initialized to point to node 0, the dummy internal node and first leaf. Each LOSER pointer may be initialized to point to the double node that contains it. This is how the LOSER pointers in Fig. 2.19 were set.

ANALYSIS OF ALGORITHM 2.19

How much time does the run construction algorithm take? Observe that the amount of time used is proportional to the number of comparisons of keys done in the loop that updates the LOSER pointers. Each time the structure is updated, a comparison is done

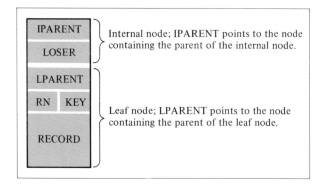

Figure 2.20
A node for replacement selection.

for each LOSER pointer on the path from a leaf to the root. Let m_0 be the number of records that can fit in memory, along with the tree structure and program, at one time. Then m_0 is the number of leaves in the tree and the number of comparisons done for each update is $\lceil \log m_0 \rceil$ or $\lfloor \log m_0 \rfloor$. (The leaves may be on two levels as in Fig. 2.18.) Suppose that there are n records on the input tape. There are m_0 dummy keys for run number 0. The structure is updated after the output step for each of these (even though no writing is done for run 0). The algorithm terminates when it encounters a dummy key from run RUNS + 1 at the output step, so the time required for the run construction algorithm is proportional to $(n + m_0) \log m_0$. The time required by the first method we considered for run construction — reading in as many records as will fit, sorting them, and writing out a run — is $O(n \log m)$, where m is the number of records that fit in memory along with the sorting program. If m and m_0 are approximately equal (and we will say more about that below), the two methods are similar timewise. (For both methods we did not take into account the time required for input and output since it would be highly dependent on the sophistication of the system on which the algorithms are implemented.)

The point of studying the replacement selection method was not to decrease the time required for run construction, but rather to increase the size of the runs generated. That, we saw, would decrease the amount of time required for the merge phase of an external sort. Have we succeeded? In the example used in this section (Figs. 2.17 and 2.18) $m_0 = 6$ but the runs constructed had lengths 7 and 9. It is possible to prove that with random keys the average run length produced by replacement selection is $2m_0$. (See the notes and references at the end of the chapter.) In practice, the records to be sorted are often already partly in order and even longer runs may be produced, but for the purpose of comparing the run construction methods we will use the length $2m_0$ for replacement selection. The exact values of m and $2m_0$ depend in part on the programs used, so our comments will not be quantitatively precise. However, we can draw one important conclusion: When the size of the records is very large relative to the size of the keys, m and m_0 are fairly close and replacement selection produces runs about twice as long as run construction by sorting. With large records m and m_0 are fairly close because the space needed for the LOSER pointers will not significantly affect the total number of records that can fit in memory at once. The sorting algorithm may use extra pointers also if the records are large. (On the other hand, if the records contained only the keys, then m_0 could be approximately equal to $\frac{1}{2}m$ because the number of nodes in the tree structure used in replacement selection is twice the number of leaves, i.e., records! In this case the two methods would produce runs of comparable sizes.) The advantage of using replacement selection increases with the size of the records.

AN APPLICATION: PRIORITY QUEUES

A priority queue is a queue in which each item has an assigned priority number and the usual first in, first out rule is modified so that an item is not removed if there are any items with higher priority still in the queue. Priority queues are often used for programming jobs submitted to a computer.

The data structure and updating algorithm for replacement selection can be used for a priority queue with priority numbers playing the role of run numbers and sequentially assigned item (or job) numbers playing the role of keys.

2.10   *Exercises*

Section 2.1: Introduction

2.1. Probably the easiest sorting algorithm to understand is one which we call MAXSORT and which works as follows: Find the largest key, say K, in the unsorted section of the file (initially the whole file) and then interchange K with the key in the last position in the unsorted section. Now K is considered part of the sorted section consisting of larger keys at the end of the file; it is no longer in the unsorted section. Repeat this procedure until the whole file is sorted.

a) Write out an algorithm for MAXSORT assuming an array L contains n keys to be sorted.

b) How many comparisons of keys does MAXSORT do in the worst case? on the average?

2.2. Consider the following outline for a sorting algorithm:

a) Find the largest (M) and second largest (S) entry in the file by using the tournament method of Chapter 1.

b) Interchange M and S with the entries in the last two positions in the file.

c) Decrement the size by 2 and repeat this procedure until the file is sorted. How many comparisons would this algorithm do in the worst case? How many comparisons would it do if the input file were already sorted?

Section 2.2: BUBBLESORT

2.3. The correctness of the BUBBLESORT algorithm depends on several claims that were made in the discussion of the algorithm. These are easy to verify but worth doing in order to consciously recognize the mathematical properties involved.

    a) Prove that after one pass through the file, the largest entry will be at the bottom.

    b) Prove that if there is no pair of consecutive entries out of order, then the entire file is sorted.

    c) Prove that if the last exchange made in some pass occurs at the jth and $(j + 1)$st positions, then all entries from the $(j + 1)$st to the nth are in their correct position. (Note that this is stronger than saying simply that these items are in order.)

2.4.   How may the keys be arranged for a worst case for **BUBBLESORT**? What inputs are "best cases" for **BUBBLESORT**, i.e., for what inputs do Algorithms 2.1 and 2.2 do the fewest comparisons? How many comparisons do they do in the best case?

2.5.   We modified **BUBBLESORT** to avoid unnecessary comparisons in the tail of the file. Can something similar be done to avoid unnecessary comparisons when the keys at the beginning of the file are already in order? If so, write out the modifications to the algorithm. If not, explain why not.

2.6.   Describe a method (or write an algorithm) to solve the following problem: Given a sequence $B = (b_1, b_2, \ldots, b_n)$ such that $0 \leqslant b_i \leqslant i - 1$ for $1 \leqslant i \leqslant n$, find the permutation σ such that $B = B(\sigma)$. ($B(\sigma)$ is defined in the discussion of the average behavior of **BUBBLESORT**.)

2.7.   Show that a permutation on n letters has at most $n(n - 1)/2$ inversions. Which permutation(s) have exactly $n(n - 1)/2$ inversions?

Section 2.3: QUICKSORT

2.8.   Write out an algorithm for **SPLIT**.

2.9.   Show by an example that the part of a file rearranged by the **SPLIT** algorithm described at the beginning of Section 2.3 may have more inversions after being split than it had before. (See the discussion of the average behavior of **BUBBLESORT** for the definition of an inversion.)

2.10. How should the keys in a file be arranged so that each time **SPLIT** is executed $L(k)$ contains the largest entry in the subfile $L(k), \ldots, L(m)$? (Be careful; your first guess is probably wrong.) How many times are keys moved from one location to another in this case?

2.11. How many comparisons does the basic **QUICKSORT** algorithm do if the file is already sorted? How many moves does it do?

2.12. Describe in a sentence or two what the basic **QUICKSORT** algorithm does if the keys in the input file are in decreasing order. How many moves are done?

2.13. How would the action of **QUICKSORT** differ if a queue were used instead of a stack to store the borders of the subfiles yet to be sorted? Consider correctness, time, and space usage.

2.14. Suppose that instead of choosing $L(k)$ as X, **SPLIT** sets X to the median of $L(k)$, $L((k + m)/2)$, and $L(m)$. How many comparisons will **QUICKSORT** do in the worst case to sort n keys? (Remember to count the comparisons done in choosing X.)

2.15. Suppose you are given an unsorted file containing n keys. Your problem is to find the tth smallest key where $1 \leqslant t \leqslant n$. QUICKSORT can be modified to do this so that in most cases it does much less work than is needed to completely sort the file. Write a modified QUICKSORT algorithm called FINDTTH for this purpose.

Section 2.4: Lower Bounds for Sorting by Comparison of Keys

2.16. Devise an algorithm to sort five keys that is optimal in the worst case.

Section 2.5: HEAPSORT

2.17. Suppose the nodes of a heap are stored in an array level by level beginning with the root and left to right within each level. Show that the left child of the node in the ith cell is in the $2i$th cell.

2.18. Suppose a file of distinct keys in decreasing order is to be sorted (into increasing order) by HEAPSORT.
   a) How many comparisons of keys are done in the heap construction phase if there are ten keys?
   b) How many are done if there are n keys? Show how you derive your answer.
   c) Is a file in decreasing order a best case, worst case, or intermediate case for the heap construction? Justify your answer.

2.19. An earlier version of HEAPSORT used the following scheme for constructing the heap: Keys are inserted one at a time by adding a new node in the next position at the bottom level and letting the new key filter up from there to its proper place. The insertion routine is:

   INSERT (K, i)
   K is the key to be inserted; i is the number of nodes in the tree after K is added.
   1. Set P to the location of the ith node
   2. **while** $P \neq$ ROOT and $K >$ KEY(PARENT(P)) **do**
   3.   KEY(P) $\leftarrow$ KEY(PARENT(P))
   4. $P \leftarrow$ PARENT(P)
      end
   5. KEY(P) $\leftarrow K$

   a) How many comparisons of keys are done in the worst case to insert the ith key?
   b) How many comparisons are done in the worst case to construct the heap using INSERT repeatedly for i from 1 to n?
   c) How many comparisons would be done in the worst case by HEAPSORT if INSERT were used to construct the heap?

Section 2.7: Bucket Sorts

2.20. Suppose RADIX SORT does m distribution passes on keys with w bits (where m is a divisor of w) and there is one bucket for each pattern of w/m bits, hence $k = 2^{w/m}$. Since mn key distributions are done, it may seem advantageous to decrease m. How large must the new k be if m is halved?

Section 2.8: Merging Sorted Lists

2.21. Prove that $BM(1, n) = 1 + \lfloor \log n \rfloor$ without using the recurrence relation Eq. (2.1). Then verify that this formula satisfies the recurrence relation and the closed form (Eq. 2.5).

2.22. Verify that $n - 2^t \geqslant m$ for $m > 1$, where $2^t m \leqslant n < 2^{t+1} m$.

2.23. a) Devise an algorithm to sort a file by merging subfiles that were sorted recursively by the same method.
   b) How many comparisons does your algorithm do?

Section 2.9: External Sorting

2.24. Suppose that the r runs constructed during the first phase of an external sort are distributed so that one run is on one tape and all the others are on another tape. How many passes over the keys will be done in the merge phase using three tapes?

2.25. Prove Lemma 2.15.

2.26. Suppose that you have a large unsorted file on a tape and only two tape drives. Carefully outline a method for sorting the file.

Additional Problems

2.27. Outline a reasonable method of solving each of the following problems.
   a) You are given a pile of thousands of telephone bills and thousands of checks sent in to pay the bills. (Assume telephone numbers are on the checks.) Find out who didn't pay.
   b) You are given a list containing title, author, call number, and publisher of all the books in a school library and another list of 30 publishers. Find out how many of the books were published by each of those 30 companies.
   c) You are given all of the book checkout cards used in the campus library during the past year. Determine how many distinct people checked out at least one book.
   d) Rearrange a file so that all the negative keys precede all the positive keys.

2.28. A sorting method is *stable* if equal keys remain in the same relative order in the sorted file as they were in the original file. (That is, a sort is stable if for any $i < j$ such that initially $L(i) = L(j)$, the sort moves $L(i)$ to $L(k)$ and moves $L(j)$ to $L(\ell)$ for some k and ℓ such that $k < \ell$.) Which of the following algorithms are stable? For each that is not, give an example in which the relative order of two equal keys is changed.
   a) BUBBLESORT
   b) QUICKSORT (If your answer depends on the details of the SPLIT algorithm, which is not given in the text, describe how those details affect your answer.)
   c) HEAPSORT
   d) SHELLSORT
   e) RADIX SORT
   f) The external sort in Section 2.9 using replacement selection

2.29. Suppose you have a file of 1000 records in which only a few are out of order and they are not very far from their correct position. If your measure of work is the number of comparisons and the number of moves done, which sorting algorithm would you use to put the whole file in order? Justify your choice.

2.30. Consider the following sorting method:

> Beginning with a sorted subfile consisting of the first key, insert the next key in its correct position by doing a binary search in the sorted subfile.

What data structure would you use to implement this algorithm? How many comparisons would be done in the worst case? What is the order of the worst-case running time?

2.31. Suppose that you have a computer with n memory locations, numbered 1 through n, and one instruction CEX, called "compare-exchange." For $1 \leqslant i, j \leqslant n$, CEX i, j compares the keys in memory cells i and j and interchanges them if necessary so that the smaller key is in the cell with the smaller index. The CEX instruction can be used to sort keys. For example, the following program sorts for $n = 3$:

$$\text{CEX } 1,2$$
$$\text{CEX } 2,3$$
$$\text{CEX } 1,2$$

a) Write an efficient program using only CEX instructions to sort six keys. (*Suggestion*: Write programs for $n = 4$ and $n = 5$ first. It is easy to write programs for $n = 4$, 5, and 6 using 6, 10, and 15 instructions, respectively. However, none of these is optimal.)

b) Write a CEX program to sort n keys in n cells for a fixed but arbitrary n. Use as few instructions as you can. Describe the strategy your program uses and include comments where appropriate. Since there are no loop and test instructions, you may use an ellipsis to indicate repetition of instructions of a certain form; for example:

$$\text{CEX } 1,2$$
$$\text{CEX } 2,3$$
$$\vdots$$
$$\text{CEX } n - 1, n$$

c) How many CEX instructions does your program for (b) have?

d) Give a lower bound on the number of CEX instructions needed to sort n keys.

2.32. Suppose you have a computer with a small memory and you are given a list of keys on a tape. Keys may be read into the CPU for processing, but no key may be read more than once.

a) What is the minimum number of storage cells needed in the CPU (aside from the program) to find the largest key in the file? Justify your answer.

b) What is the minimum number of cells needed to find the median? Justify your answer.

2.33. a) Write an algorithm which, when given an array L of records with keys a_1, $a_2, \ldots, a_n$ and a permutation π of the numbers $1, 2, \ldots, n$, rearranges the records in the order $a_{\pi(1)}, a_{\pi(2)}, \ldots, a_{\pi(n)}$. You may assume the values of π are given in another array. Assume that the records in L are large; in particular, they will not fit in the π array. Your algorithm may destory π. If you use extra space, state how much.

b) What is the total number of moves done by your algorithm in the worst case? Is the running time of your algorithm proportional to the number of moves? If so, explain why. If not, what is the order of the running time?

2.34. Give a lower bound for the number of comparisons necessary to find the median in an unsorted list of n keys. For simplicity you may assume n is odd and the keys are distinct. The median is the $[(n + 1)/2]$ th largest key.

2.35. Let f be a function defined on the interval from 0 to n, i.e., for $0 \leqslant x \leqslant n$, which achieves its maximum value at x_M, a point in the interval, and is strictly increasing for $0 \leqslant x < x_M$ and strictly decreasing for $x_M < x \leqslant n$. (*Note:* x_M may be 0 or n.)

a) The only operation you can do with f is evaluate it at points in the interval. Write an algorithm to find x_M with an error of at most ϵ, where $0 < \epsilon < 1$; i.e., if your algorithm's answer is $\bar{x}$, then $|\bar{x} - x_M| < \epsilon$. Use instructions of the form $y \leftarrow f(x)$ to indicate evaluation of f at the point x.

b) How many evaluations does your algorithm do in the worst case? (You should be able to devise an algorithm that is better than linear in n.)

PROGRAMS

For each program include a counter that counts comparisons of keys. Include among your test data files in which the keys are in decreasing order, increasing order, and random order. Use files of several sizes. Output should include the number of keys and the number of comparisons done.

1. QUICKSORT: Use the improvements described at the end of Section 2.3.

2. HEAPSORT: Print out the full heap after all the keys have been inserted.

3. RADIX SORT

4. Run construction by replacement selection: The run construction algorithm may be implemented and tested with the "memory size" set to, say, 100. The runs constructed may be printed out on the line printer. Data cards may be used in place of input tapes but since several hundred should be used, a better alternative is to include a subprogram that generates random keys.

5. If tapes or disks are available for student use, implement a complete external sort.

6. BINARY MERGE

NOTES AND REFERENCES

Much of the material in this chapter is based on Knuth (1973), without a doubt the major reference on sorting and related problems. The interested reader is strongly encouraged to consult this book for more algorithms, analysis, exercises, and references.

Page and section numbers in the following references refer to Knuth (1973) unless otherwise indicated. The average number of inversions in a permutation (our Section 2.2.: BUBBLESORT) is derived from material in Knuth's Section 5.1.1 on inversions. See also pp. 82 and 108–109. At the end of Section 2.5 we commented that there are algorithms that do fewer comparisons than HEAPSORT in the worst case. The Ford-Johnson algorithm, called MERGE INSERTION (pp. 185–188), is one such. It is known to be optimal for small values of n. BINARY INSERTION (p. 83) is another algorithm that does approximately $n \log n$ comparisons in the worst case. A discussion of various choices of increments for use in SHELLSORT may be found in pp. 90–95. Theorem 2.10 is there also. The formula in Lemma 2.14 of Section 2.9 is derived with the help of generating functions. See Knuth (1969), Section 1.2.8, for this and other properties of the Fibonacci numbers that are useful in the analysis of algorithms. The generalization of the Fibonacci distribution for a large number of tapes is discussed in Knuth (1973) in the section on the polyphase merge (pp. 266–273). See pp. 254–256 and 260–262 for the size of the runs produced by replacement selection, and Section 5.4.9 for the nitty-gritty details related to the use of disks and drums for sorting.

Many sorting algorithms are known by more than one name and it will aid the reader who consults other references to be aware of the alternative names. BUBBLE-SORT is sometimes called EXCHANGE SORT. SHELLSORT is referred to as "sorting by diminishing increments." (The algorithm originated with Donald Shell, hence the name used in this book.) QUICKSORT is also known as partition-exchange sort.

Graphs and Digraphs

3

3.1 *Definitions and Representations*

Informally, a graph is a finite set of points, some of which are connected by lines, and a digraph (short for "directed graph") is a finite set of points, some of which are connected by arrows. Graphs and digraphs are useful abstractions for numerous problems and structures in operations research, computer science, electrical engineering, economics, mathematics, physics, chemistry, communications, game theory, and many other areas. Consider the following examples.

Example 3.1 A (hypothetical) map of airline routes between several California cities

The points are the cities; a line connects two cities if and only if there is a nonstop flight between them in both directions. See Fig. 3.1.

Example 3.2 The flow of control in a flowchart

The points are the flowchart boxes; the connecting arrows are the flowchart arrows. See Fig. 3.2.

Example 3.3 A binary relation

Let S be the set $\{1, 2, \ldots, 9, 10\}$ and let R be the relation on S defined by xRy if and only if $x \neq y$ and x divides y. In the digraph in Fig. 3.3 the points are the elements of S and there is an arrow from x to y if and only if xRy.

Example 3.4 An electrical circuit

The points could be diodes, transistors, capacitors, switches, etc. Two points would be connected by a line if there is a wire connecting them.

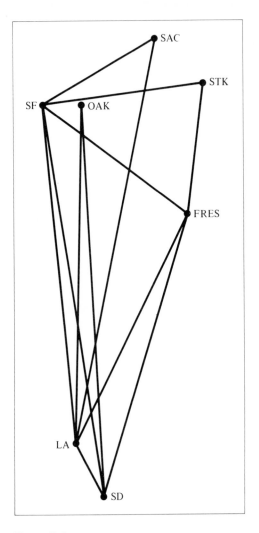

Figure 3.1
A graph of nonstop airline flights between
California cities.

Formally, a *graph*, G, is a pair (V, E) where V is a finite set whose elements are called vertices and E is a set of subsets of V of order two. Elements of E are called edges.

For the graph in Fig. 3.1, for example, we have $V = \{SF, OAK, SAC, STK, FRES, LA, SD\}$ and $E = \{\{SF, STK\}, \{SF, SAC\}, \{SF, LA\}, \{SF, SD\}, \{SF, FRES\}, \{SD, OAK\}, \{SAC, LA\}, \{LA, OAK\}, \{LA, FRES\}, \{LA, SD\}, \{FRES, STK\}, \{SD, FRES\}\}$.

A *digraph, G,* is a pair (V, E) where V is a finite set whose elements are called vertices, and E is a set of ordered pairs of distinct elements of V. Elements of E are called edges. For (v, w) in E, v is the *tail* and w the *head* of (v, w).

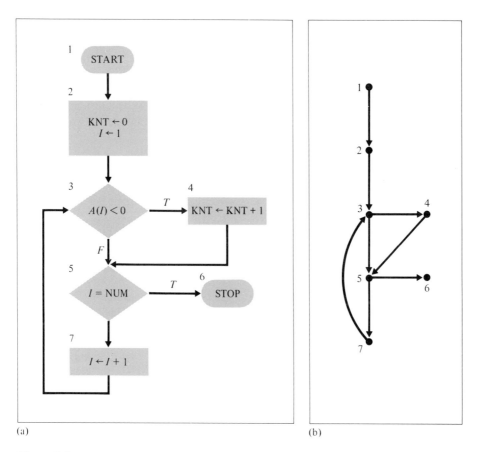

(a)   (b)

Figure 3.2
(a) A flowchart, (b) A directed graph. Arrowheads indicate the direction of flow.

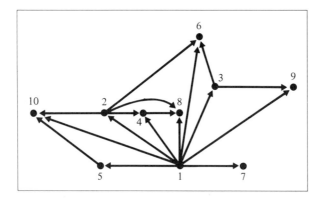

Figure 3.3
The relation R in Example 3.3.

In the flowchart example (Fig. 3.2b), using the numbers assigned to the flowchart boxes, $V = \{1, 2, \ldots, 7\}$ and $E = \{(1, 2), (2, 3), (3, 4), (3, 5), (4, 5), (5, 6), (5, 7), (7, 3)\}$.

The definitions of graph and digraph imply that there cannot be an edge that connects a vertex to itself and that there cannot be two edges between one pair of vertices in a graph, or two edges with the same orientation (direction), betwe. n one pair of edges in a digraph. In some applications these restrictions may be dropped. For example, let R be a binary relation on a finite set S and let $V = S$. We may wish to have an edge (x, y) for each pair x and y in S such that xRy, even if $x = y$. If the definitions of graph and digraph are modified in some application to permit such edges, care must be taken to see that any theorems and algorithms used for graphs and digraphs are still correct. In this chapter all graphs and digraphs will satisfy the definitions given.

The four examples above should be sufficient to illustrate that graphs and digraphs provide a natural abstraction of relationships of diverse objects, including both physical objects and their arrangement such as cities connected by airline routes, highways, or railway lines, and abstract things such as binary relations and the control structure of a program. These examples should also suggest some of the questions we may wish to ask about the objects represented, questions that will be rephrased in terms of the graph or digraph and can be answered by algorithms that work on these structures. The question "Is there a nonstop flight betwen San Diego and Sacramento?" translates into "Is there an edge between the vertices SD and SAC in Fig. 3.1?" Consider the following questions:

What is the cheapest way to fly from Stockton to San Diego?

Which route involves the least flying time?

If one city's airport is closed by bad weather, can you still fly between any other pair of cities?

How much traffic can flow from one specified point to another using certain specified roads?

Is a given binary relation transitive?

Does a given flowchart have any loops?

How should wires be attached to various electrical outlets so that all are connected together using the least amount of wire?

In this chapter we will study algorithms to answer most of these questions. The remainder of this section is devoted to definitions, general remarks, and a discussion of data structures useful for representing graphs and digraphs in a computer. Many statements and definitions are applicable to both graphs and digraphs and we will use a common notation for both to minimize repetition. An edge $\{v, w\}$ in a graph or (v, w) in a digraph will be written $\overline{vw}$. (For a graph, of course, $\overline{vw} = \overline{wv}$.)

A *subgraph* of a graph or digraph $G = (V, E)$ is a graph (or digraph) $G' = (V', E')$ such that $V' \subseteq V$ and $E' \subseteq E$. A *complete graph* is a graph with an edge between each pair of vertices. Vertices v and w are said to be *incident* with the edge $\overline{vw}$ and vice versa.

The edges of a graph or digraph $G = (V, E)$ induce a relation called the adjacency relation, A, on the set of vertices. Let v and w be elements of V. Then wAv (read "w is *adjacent* to v") if and only if $\overline{vw} \in E$. In other words, wAv means w can be reached from v by moving along an edge of G. If G is a graph, the relation A is symmetric.

Consider Fig. 3.1 again and suppose we wish to travel by airplane from Los Angeles (LA) to Fresno (FRES). There is an edge $\overline{\text{LA FRES}}$ that is one possible route, but

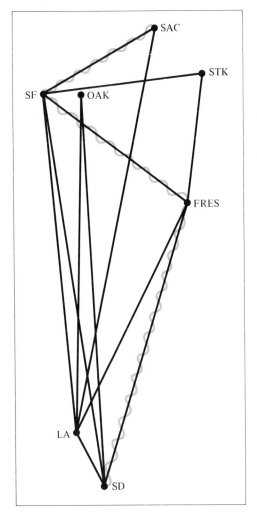

Figure 3.4
A path from SD to SAC.

there are others. We could go from LA to SAC to SF to FRES, or we could go from LA to SD to FRES. These are all "paths" from LA to FRES in the graph. The concept of a path is very useful in many applications, some (like this one) that involve routing of people, telephone messages, automobile traffic, liquids or gases in pipes, etc., and others where paths represent abstract properties. (See Exercise 3.3.) Formally a *path from v to w* in a graph or digraph $G = (V, E)$ is a sequence of edges $\overline{v_0 v_1}, \overline{v_1 v_2}, \ldots,$ $\overline{v_{i-1} v_i}, \overline{v_i v_{i+1}}, \ldots, \overline{v_{k-1} v_k}$, such that $v_0 = v$, $v_k = w$, and $v_0, v_1, \ldots, v_k$ are all distinct. The length of the path is k. A vertex v alone is considered to be a path of length zero from v to itself. The path $\overline{SD\ FRES}, \overline{FRES\ SF}, \overline{SF\ SAC}$ is shown in Fig. 3.4. We will denote a path by simply listing the vertices through which it passes. A graph is *connected* if for each pair of vertices, v and w, there is a path from v to w. Connectedness for digraphs may be defined in either of two ways, depending on whether or not we require that edges be traversed only from tail to head. Thus we have the following definitions. A digraph $G = (V, E)$ is *weakly connected* if, for each pair of vertices v and w, there is a sequence of vertices $v_0, v_1, \ldots, v_k$ such that $v_0 = v$, $v_k = w$, and for $i = 0, 1, \ldots, k - 1$ either $\overline{v_i v_{i+1}} \in E$ or $\overline{v_{i+1} v_i} \in E$. G is *strongly connected* if, for each pair v and w, there is a path from v to w.

A *cycle* in a graph or digraph $G = (V, E)$ is like a path, $v_0, v_1, \ldots, v_k$, except that $v_k = v_0$. A graph or digraph is *acyclic* if it has no cycles. A *tree* may be defined as a connected, acyclic graph; see Fig. 3.5. A *rooted tree* is a tree with one vertex singled out as the root. The parent and child relations often used with trees can be derived once a root is specified.

In many applications of graphs and digraphs it is useful or necessary to associate a number with each edge. The numbers represent costs or benefits derived from using the particular edge in some way. Consider, once again, Fig. 3.1 and suppose that we want to fly from SD to SAC. There is no nonstop flight, but there are several routes or paths that could be used. Which one is best? To answer this question we need a standard by which to judge the various paths. Some possible standards are:

1. The number of stops,

2. The total ticket cost, or

3. The total flying time.

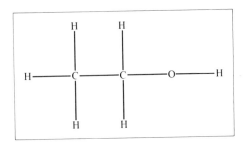

Figure 3.5
A tree: an alcohol molecule.

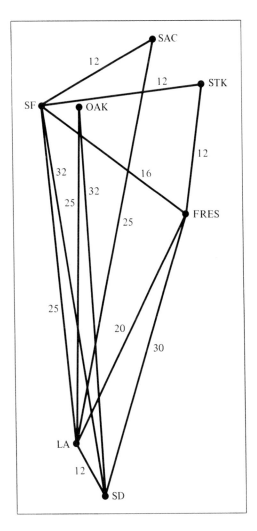

Figure 3.6
A weighted graph showing airline fares.

After choosing a standard, we could assign to each edge in the graph the cost (in stops, money, or time) of traveling along that edge. The total cost of a particular path is the sum of the costs of the edges traversed by that route. Figure 3.6 shows the airline route graph with the approximate cost of a plane ticket written beside each edge. The reader may verify that the cheapest way to get from **SD** to **SAC** is to make one stop in **LA**. The general problem of finding "best" paths is studied in Section 3.3.

Figure 3.7 shows some of the streets in a city and might be used to study the flow of automobile traffic. The number assigned to an edge indicates the amount of traffic that can flow along that section of the street in a certain time interval. The

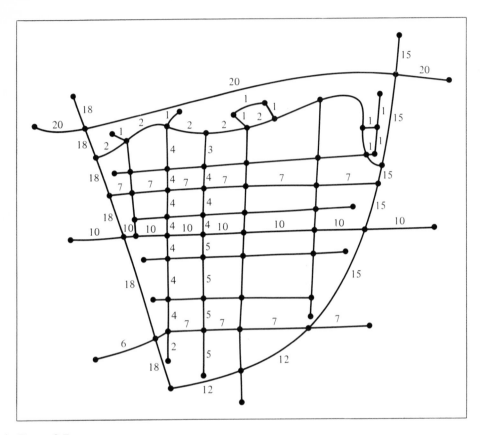

Figure 3.7
A street map showing traffic capacities.

number is determined by the type and size of road, the speed limit, the number of traffic lights between the intersections shown in the graph as vertices (assuming not every street is shown in the graph), and various other factors.

The assignment of numbers to edges occurs often enough in applications to merit a definition. A *weighted graph* (or *weighted digraph*) is a triple (V, E, W) where (V, E) is a graph (or digraph) and W is a function from E into $\mathbf{Z}^+$, the positive integers. If $e \in E$, $W(e)$ is called the *weight* of the edge e. The weights in some applications will correspond to costs or undesirable aspects of an edge, whereas in other problems the weights are capacities or similar beneficial properties of the edges. Sometimes it is desirable to allow negative or noninteger weights, but one must be careful when choosing algorithms since the correctness of some of them depends on restricting the weights to nonnegative integers.

COMPUTER REPRESENTATION OF GRAPHS AND DIGRAPHS

We have seen two ways of representing a graph or digraph on paper: by drawing a picture in which vertices are represented by points and edges as lines, and by listing the vertices and edges. For solving problems on graphs in a computer, we need other representations. Let $G = (V, E)$ be a graph or digraph with $|V| = n$, $|E| = m$, and $V = \{v_1, v_2, \ldots, v_n\}$.

G can be represented by an $n \times n$ matrix $A = (a_{ij})$, called the *adjacency matrix* for G. A is defined by

$$a_{ij} = \begin{cases} 1 & \text{if } \overline{v_i v_j} \in E \\ 0 & \text{otherwise} \end{cases} \quad \text{for } 1 \leqslant i, j \leqslant n.$$

The adjacency matrix for a graph is symmetric and only half of it need be stored. If $G = (V, E, W)$ is a weighted graph or digraph, the weights can be stored in the adjacency matrix by modifying its definition as follows:

$$a_{ij} = \begin{cases} W(\overline{v_i v_j}) & \text{if } \overline{v_i v_j} \in E \\ c & \text{otherwise} \end{cases} \quad \text{for } 1 \leqslant i, j \leqslant n,$$

where c is a constant whose value depends on the interpretation of the weights and the problem to be solved. If the weights are thought of as costs, ∞ (or some very high number) may be chosen for c because the cost of traversing a nonexistent edge is prohibitively high. If the weights are capacities, a choice of $c = 0$ is usually appropriate since nothing can move along an edge that is not there.

Algorithms for solving some problems on graphs and digraphs require that every edge be examined and processed in some way at least once. If an adjacency matrix representation is used, we may as well think of a graph or digraph as having edges between all pairs of distinct vertices, though some have weight 0 or ∞, because algorithms may have to examine each entry in the matrix to determine which edges really exist. Since the number of edges is $n(n - 1)/2$, or $n(n - 1)$ in a digraph, the complexity of such algorithms will be at least $\Theta(n^2)$.

An alternative to the adjacency matrix representation is a data structure containing, for each vertex v, a linked list of nodes indicating which vertices are adjacent to v. The data in the adjacency lists will vary with the problem but there is a fairly standard basic structure that is useful for many algorithms. Suppose we number the vertices so that $V = \{1, 2, \ldots, n\}$. We use an array ADJLIST containing n listheads, one for the adjacency list of each vertex. The nodes in the linked lists have the form

VTX	LINK

where VTX is a vertex number and LINK is a pointer field. Such a node represents an edge in the graph or digraph. In particular if

2	LINK

is on the adjacency list for 7, it represents the edge $\overline{72}$. This data structure for a graph is illustrated by the example in Fig. 3.8. Each edge is represented twice; that is, if $\overline{vw} \in E$ there is a node for w on the adjacency list for v and a node for v on the adjacency list for w. Thus there are $2m$ linked list nodes and n listheads. For a digraph each edge is represented exactly once. If the graph or digraph is weighted, a weight

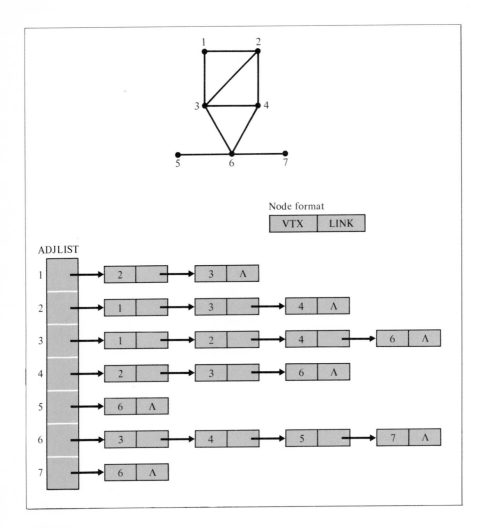

Figure 3.8
Adjacency list structure for a graph.

field is included in each node. Figure 3.9 illustrates the structure for a weighted digraph. The form of the list nodes is

VTX	WGT	LINK

Additional fields may be added to the list heads or the linked list nodes as required by the algorithms to be used. Note that if the nodes within an adjacency list appear in a different order, the structure still represents the same graph or digraph, but an algorithm using the list will encounter the nodes in a different order and may behave very differently. An algorithm should not assume any particular ordering.

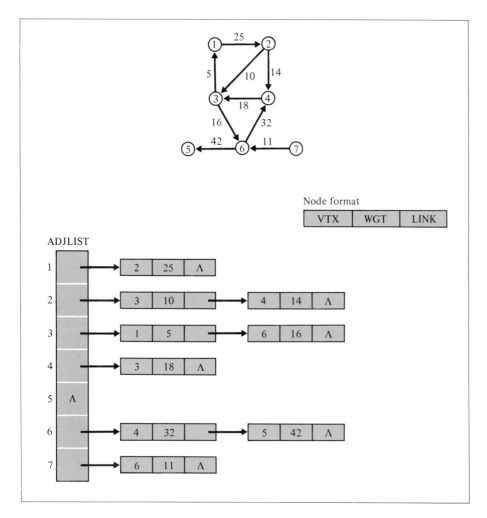

Figure 3.9
Adjacency list structure for a weighted digraph.

3.2   *A Minimal Spanning Tree Algorithm*

The first problem we will study is the problem of finding a minimal spanning tree for a weighted graph (not directed).

A *spanning tree* for a graph, $G = (V, E)$, is a subgraph of G which is a tree and contains all the vertices of G. In a weighted graph the weight of a subgraph is the sum of the weights of the edges in the subgraph. A *minimal spanning tree* for a weighted graph is a spanning tree with minimal weight. There are many situations in which minimal spanning trees must be found. Whenever one wants to find the cheapest way to connect a set of terminals, be they cities, electrical terminals, computers, or factories, by using say road, wires, or telephone lines, a solution is a minimal spanning tree in the graph with an edge for each possible connection weighted by the cost of that connection. A weighted graph may have more than one minimal spanning tree.

The algorithm presented in this section begins by selecting an arbitrary node, and then "branches out" from the part of the tree constructed so far by choosing an edge of minimal weight that leads to a new vertex. During the course of the algorithm the vertices and edges may be thought of as being divided into three (disjoint) sets each. The sets are as follows.

V1   Vertices that are in the part of the spanning tree already constructed

E1   Edges that have been chosen so far (they form a minimal spanning tree for vertices in V1)

V2   Vertices that are not in V1 but are adjacent to some vertex in V1

E2   Contains, for each vertex v in V2, an edge of minimal weight connecting some vertex in V1 with v (hence there is a one-to-one correspondence between vertices in V2 and edges in E2)

V3   The remaining vertices

E3   The remaining edges

The algorithm will halt when all of the vertices are in V1. Figure 3.10 shows what might be in these sets at some intermediate point in the algorithm. Given the sets as described, the next step is to pull one more vertex into V1 by choosing from E2 an edge with minimal weight — call it $\overline{xy}$, where $x \in$ V1 and $y \in$ V2 — and putting it in E1 and y in V1.

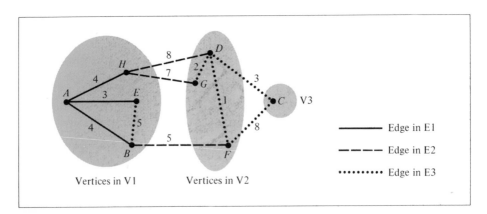

Figure 3.10
Sets V1, V2, V3, E1, E2, and E3 for the minimal spanning tree algorithm.

Can we be sure that this procedure will yield a minimal spanning tree? Since the algorithm will start with E1 empty, we may assume that the edges already in E1 form part of a minimal spanning tree and prove that after adding to it a minimal edge from E2, E1 is still contained in a minimal spanning tree. (Note that we refer to a minimal spanning tree as if it were simply a set of edges. The vertices in the tree are those incident with the edges and need not be explicitly mentioned.)

Theorem 3.1 Let $G = (V, E, W)$ be a weighted graph and let $El \subseteq E$ be a subset of the edges in some minimal spanning tree T for G. Let V1 be the vertices incident with edges in E1. If $\overline{xy}$ is an edge of minimal weight such that $x \in V1$ and $y \notin V1$, then $El \cup \{\overline{xy}\}$ is a subset of a minimal spanning tree.

Proof. If $\overline{xy}$ is in T, the conclusion follows. Suppose $\overline{xy}$ is not in T. There is a path from x to y in T since trees are connected. Let $\overline{vw}$ be the first edge in that path with exactly one vertex in V1, say v. (See Fig. 3.11.) Let $T' = T - \{\overline{vw}\} \cup \{\overline{xy}\}$. The reader may verify that $El \cup \{\overline{xy}\} \subseteq T'$ and T' is a spanning tree. Since v is in V1 and w is not in V1, by the choice of $\overline{xy}$, $W(\overline{xy}) \leqslant W(\overline{vw})$ so $W(T') \leqslant W(T)$ and T' is a minimal spanning tree. $\square$

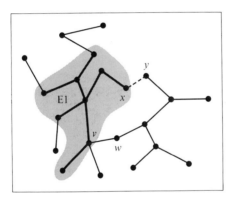

Figure 3.11
Minimal spanning tree T for Theorem 3.1. Edges shown as solid lines are in T. The heavy edges are also in E1. There may be many other edges in the graph that are not shown.

After moving the new edge $\overline{xy}$ from E2 to E1, the conditions that define V2 and E2 must be reestablished. In Fig. 3.10, the edge chosen from E2 is $\overline{BF}$. $\overline{HD}$ is in E2 and has weight 8, but after F is put in V1, the edge $\overline{FD}$, with weight 1, should replace $\overline{HD}$ in E2. Also, when F is in V1, the vertex C must be put in V2 since it is now adjacent to a member of V1. The edge $\overline{FC}$ goes in E2. The algorithm makes such adjustments in the **for** loops in steps 3 and 4 below.

We can give a complete description now of how the algorithm manipulates the sets V1, V2, V3, E1, and E2. (E3 need not be explicitly used.) Later we will consider data structures and implementation details and rewrite the algorithm to conform more closely to a way in which it may be implemented.

Algorithm 3.1 MINIMAL SPANNING TREE

Input: $G = (V, E, W)$, a weighted graph.

Output: E1, the edges in a minimal spanning tree.

1.   [Initialize the sets.]
    Let x be an arbitrary vertex. V1 $\leftarrow$ $\{x\}$; V2 $\leftarrow$ ϕ; V3 $\leftarrow$ V $- \{x\}$; E1 $\leftarrow$ E2 $\leftarrow$ ϕ
2.   **while** V1 $\neq V$ **do**
3.   **for** each y in V2 adjacent to x
        if $W(\overline{xy}) < W$(the edge e in E2 incident with y)
        then E2 $\leftarrow$ E2 $- \{e\} \cup \{\overline{xy}\}$
4.   **for** each y in V3 adjacent to x **do**
        V2 $\leftarrow$ V2 $\cup \{y\}$; V3 $\leftarrow$ V3 $- \{y\}$
        E2 $\leftarrow$ E2 $\cup \{\overline{xy}\}$
    end
5.   **if** E2 $= \phi$ **then return** 'no spanning tree'
6.   Find an edge, e, in E2 with minimum weight;
    Set x to be the vertex in V2 incident with e.
    E1 $\leftarrow$ E1 $\cup \{e\}$; E2 $\leftarrow$ E2 $- \{e\}$
    V1 $\leftarrow$ V1 $\cup \{x\}$; V2 $\leftarrow$ V2 $- \{x\}$
    end

The algorithm terminates at line 2 and the edges in E1 form a minimal spanning tree if and only if G is connected. Otherwise it terminates at step 5 and the edges in E1 form a minimal spanning tree for a connected component of G. With slight modification, Algorithm 3.1 can be used to find a set of minimal spanning trees, one for each connected component of G. Figure 3.12 contains an example of the action of the algorithm.

Can we determine how much time this algorithm requires before we consider the details of an implementation? The question should prompt another from the reader: What basic operation should we choose to count? In steps 3 and 6 weights are compared, but in step 4 there are no comparisons, although the **for** loop there apparently processes each vertex in V3 and may do a lot of work. In several steps the sets of vertices and edges are modified. We may take the number of comparisons of weights as the measure of work done and consider the other operations as "bookkeeping," but this is acceptable only if the time spent on manipulating the sets and doing other operations is proportional to the number of comparisons. Is it? To answer this question we have to examine the bookkeeping work very carefully, but the specific operations and the number of them that would be carried out depend on how the sets of vertices and edges are represented and how such lines as "**for** each y in V2 adjacent to x" are implemented. For example, maintaining V2 as a list and testing each element in it

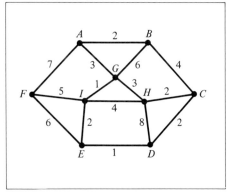

(a) A weighted graph.

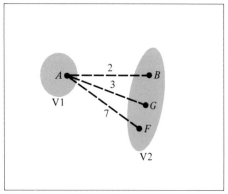

(b) After the first pass through
Steps 3 and 4.

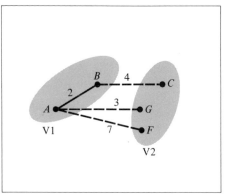

(c) $\overline{BG}$ was considered but did not
replace $\overline{AG}$ in E2.

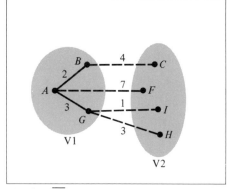

(d) After $\overline{AG}$ was selected
and Steps 3 and 4 executed.

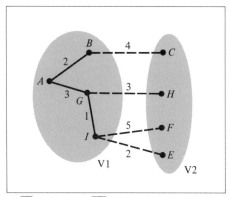

(e) $\overline{IF}$ has replaced $\overline{AF}$ in E2.

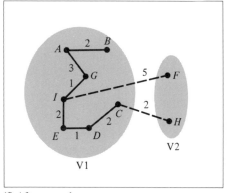

(f) After several more passes;
the two edges in E2 will be put in E1.

Figure 3.12
An example for the minimal spanning tree algorithm.

for adjacency to x may be much more, or much less, efficient than traversing an adjacency list for x and testing each element in it for membership in V2. Hence we will defer the timing analysis until after our study of data structures and implementation details. (It turns out that even with a good implementation, there are examples of graphs where the running time is of higher order than the number of comparisons done.)

IMPLEMENTATION

For efficiency we want to choose a data structure that stores only information that is really needed, and stores it in such a way that the operations required by the algorithm can be done quickly. The data structure we will use is illustrated in Fig. 3.13 and explained in the next few paragraphs. Let $n = |V|$ and $m = |E|$.

The **for** loops in steps 3 and 4 suggest that each vertex in V2 (and V3) be considered in turn and tested for adjacency to x. Suppose we use a linked list representation of these sets. Adding and deleting vertices would be easy with links. Even before considering representations for the other sets we can now derive a lower bound on the total number of operations done in steps 3 and 4. Since there are $n - 1$ vertices in V2 ∪ V3 when the algorithm starts and only one is removed (i.e., put in V1) each time through the outer loop, traversing V2 and V3 $n - 1$ times will require $\Theta(n^2)$

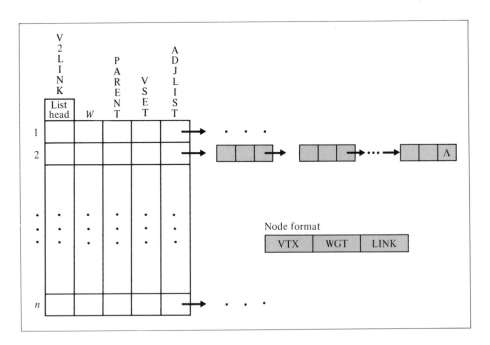

Figure 3.13
Data structure for Algorithm 3.2.

operations. This might of course be the best we can do, but let us consider the alternative mentioned earlier.

Suppose that the graph is represented by linked adjacency lists roughly as described in Section 3.1. We could implement steps 3 and 4 by traversing the adjacency list for x and testing each y on the list for membership in V2 or V3. An array VSET indexed by the vertices could be used to tell which set a particular vertex is in. That is, if $V = \{1, 2, \ldots, n\}$, VSET$(v) = 1$, 2, or 3 to indicate that v is in V1, V2, or V3. Since each vertex in V plays the role of x in the algorithm at most once, each adjacency list would be traversed at most once. The number of operations required to traverse these lists is proportional to the number of edges, m, in the graph. Although m can be as large as $(n^2 - n)/2$ (in a complete graph), for many graphs on which the algorithm may be used, it is much smaller. So although the alternative method considered for controlling the loops in steps 3 and 4 will not differ significantly for graphs with a large number of edges, the second alternative is better for some inputs and it is the one that we will use.

How should E2 be represented? We must be able to find the edge incident with a given vertex in V2 (step 3), replace edges (step 3), and find an edge with minimal weight and delete it (step 6). Our representation should permit implementation of these operations without unnecessary searching of lists or arrays. There are many that meet this criterion; we choose one that uses less space than others. Since there is one edge in E2 for each vertex in V2, we link these vertices and store data describing the corresponding edges with the links. Specifically, we use three arrays as follows. For v in V2, V2LINK(v) is the link to the next vertex in V2. V2LINK has an extra cell that serves as a listhead to facilitate insertions and deletions. PARENT(v) is the vertex that is the parent of v in the tree formed by the edges in E1 $\cup$ E2 taking the first vertex chosen (at step 1) as the root; in other words, PARENT(v) is the other vertex incident with the edge in E2 that leads to v. $W(v)$ is the weight of that edge. $W(v)$ is copied from a WGT field in an adjacency list so it can be found easily when needed in step 3. With these arrays, replacing an edge in E2 (step 3) requires only changing an entry in PARENT and W. Finding and deleting an edge in E2 with minimal weight can be done by traversing the list of vertices in V2 via the V2LINKs and comparing the corresponding entries in W. When the algorithm terminates, E1 is the set $\{z \overline{\text{PARENT}}(z)$: all z except the root$\}$, and the array W contains the weights of the selected edges. Figure 3.14 shows the data structure at an intermediate point in the execution of the algorithm on the example in Fig. 3.12.

Instead of testing "V1 $\neq V$" in step 2, which would require a scan of the VSET array, we use a counter ECOUNT for edges in E1; a tree with n vertices must have exactly $n - 1$ edges.

Algorithm 3.1 is rewritten here as Algorithm 3.2 to reflect our choice of data structure.

Algorithm 3.2 MINIMAL SPANNING TREE

Input: $G = (V, E, W)$, a weighted graph with $V = \{1, \ldots, n\}$ represented by an adjacency list structure.

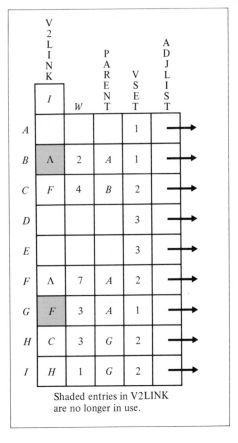

(Adjacency lists not shown. Nodes are assumed to be in alphabetical order within each list.)

	V2LINK	W	PARENT	VSET	ADJLIST
A				1	→
B	Λ	2	A	1	→
C	F	4	B	2	→
D				3	→
E				3	→
F	Λ	7	A	2	→
G	F	3	A	1	→
H	C	3	G	2	→
I	H	1	G	2	→

Shaded entries in V2LINK are no longer in use.

Figure 3.14
Minimal spanning tree data structure for the situation in Fig. 3.12(d).

Output: A list of edges in a minimal spanning tree for G.

Comment: This algorithm uses the arrays shown in Fig. 3.13 and described in the previous paragraphs.

1.   [Initialize the sets.]
        $x \leftarrow 1$; VSET(1) $\leftarrow$ 1; ECOUNT $\leftarrow$ 0; V2LINK(0) $\leftarrow \Lambda$
        for $i \leftarrow 2$ **to** n; VSET(i) $\leftarrow$ 3
2.   **while** ECOUNT $< n - 1$ **do**
3.& 4.   [Traverse the adjacency list for x.]
        PTR $\leftarrow$ ADJLIST(x)
        while PTR $\neq \Lambda$ **do**
            $y \leftarrow$ VTX(PTR)
            if VSET(y) = 2 and WGT(PTR) $< W(y)$

then do [Replace the edge to y in E2 by $\overline{xy}$.]
    PARENT$(y) \leftarrow x$; $W(y) \leftarrow$ WGT(PTR)
  end
if VSET$(y) = 3$
then do [Put y in V2 and $\overline{xy}$ in E2.]
    VSET$(y) \leftarrow 2$
    V2LINK$(y) \leftarrow$ V2LINK(0); V2LINK(0) $\leftarrow y$
    PARENT$(y) \leftarrow x$; $W(y) \leftarrow$ WGT(PTR)
  end
PTR $\leftarrow$ LINK(PTR)
end

5.   **if** V2LINK(0) = Λ **then return** 'no minimal spanning tree'.
6.   Find an edge in E2 with minimal weight; let x be the vertex in V2 incident with the edge. Remove x from V2. [In other words, traverse V2LINKs to find x such that $W(x)$ is minimal. Adjust V2LINKs to remove x.]
    VSET$(x) \leftarrow 1$; ECOUNT $\leftarrow$ ECOUNT $+ 1$
end
7.   **for** $i \leftarrow 2$ **to** n; **output** $(i,$ PARENT$(i))$

ANALYSIS (TIME AND SPACE)

Let $n = |V|$ and $m = |E|$. The number of operations done in step 1 is linear in n. We have already observed that the total amount of bookkeeping work done to traverse the adjacency lists in steps 3 and 4 is $\Theta(m)$.

The number of operations (comparisons of weights, changes in pointer values, etc.) done for each edge examined is independent of n and m, so the total time required for the work done in steps 3 and 4 is $\Theta(m)$. The test in step 5 is executed at most $n - 1$ times. So far it looks like the running time of the algorithm may be linear in m. However, as many as (roughly) $n^2/2$ comparisons may be done at step 6 even if the number of edges is smaller. E2 may contain $n - 1$ edges after the first pass through the loop and finding one with minimum weight would require $n - 2$ comparisons. Both of these numbers decrease by one on each subsequent pass, and the total number of operations is $\Theta(n^2)$. Thus the worst-case running time, as well as the worst-case number of comparisons done, is $\Theta(n^2)$. (The reader is encouraged to investigate ways of reducing the work done in step 6, but see Exercises 3.4 – 3.7.)

The data structure in Fig. 3.13 uses $4n$ cells aside from those in the adjacency list representation of the graph. This is more extra space than is used by any of the algorithms we have studied so far, and it may seem like quite a lot. However, it allows the algorithm to be implemented very efficiently timewise. (It would be worse if the extra space requirement were $\Theta(m)$ since m may be as large as $\Theta(n^2)$.)

LOWER BOUND

How much work is essential for finding a minimal spanning tree? We claim that any minimal spanning tree algorithm requires time that is at least $\Theta(m)$ in the worst case because it must examine, or process in some way, every edge in the graph. To see this, let G be a connected weighted graph where each edge has weight at least 2, and suppose there were an algorithm that did not do anything at all to an edge, $e = \overline{xy}$, in G. Then e is not in the output, T, of the algorithm. T contains a path from x to y so there is a cycle in G consisting of e and the edges in that path. Let the weight of e be 1. This could not change the action of the algorithm because it never examined e. Now, the tree obtained by removing one of the edges in the cycle and using e instead is a spanning tree with lower weight than T. The algorithm, therefore, is not correct.

3.3   *A Shortest-Path Algorithm*

In Section 3.1 we briefly considered the problem of finding the best route between two cities on a map of airline routes (Fig. 3.6). Using as our criterion the price of the plane tickets, we observed that the best — i.e., cheapest — way to get from San Diego to Sacramento was to make one stop in Los Angeles. This is one instance, or application, of a very common abstract problem on a weighted graph or digraph: finding a "shortest" path between two specified vertices. The weight of a path $v_0, v_1, \ldots, v_k$ in a weighted graph or digraph $G = (V, E, W)$ is $\sum_{i=0}^{k-1} W(\overline{v_i v_{i+1}})$, in other words, the sum of the weights of the edges in the path. If the path is called P we denote its weight by $W(P)$. A path from v to w is a *shortest path* from v to w if there is no path from v to w with lower weight. (It is, alas, conventional to mix the terminology of weight and length.) Observe that shortest paths are not necessarily unique.

How did we determine the shortest path from SD to SAC in Fig. 3.6? In fact, the reader probably used a very unalgorithmic method full of assumptions, such as that the fares are proportional to the distance between the cities and that the map is drawn approximately to scale. Then, probably, the reader picked a route that "looked" short. This is hardly an algorithm we would expect to program for a computer. We mention it to answer the above question honestly; people generally use very unrigorous ways to solve problems, especially on very small sets of data. In practice the problem of finding a shortest, or cheapest, path between two vertices in a graph or digraph arises in applications where V may contain several hundred vertices. An algorithm could consider all possible paths and compare their weights but that could take a very long time. In this section we study a shortest-path algorithm that is very similar in approach and timing to the minimal spanning tree algorithm in the previous section.

We are given a weighted graph or digraph $G = (V, E, W)$ and two specified vertices v and w; the problem is to find a shortest path from v to w. The *distance* from a vertex x to a vertex y, denoted $d(x, y)$, is the weight of a shortest path from x to y. The algorithm will find shortest paths from v to the other vertices in order of increasing distance from v. It stops when it reaches w (though it can be modified to find the set of shortest paths from v to each other vertex if desired). The algorithm, like the one in Section 3.2, starts at one vertex (v) and "branches out" by selecting certain edges that lead to new vertices. The vertices and edges are again divided into three sets each, V1, V2, V3, and E1, E2, and E3. The defining characteristics for the sets here are as follows.

V1   Vertices for which a shortest path from v has been found

E1   Edges in these paths

V2   Vertices that are not in V1 but that are adjacent to some vertex in V1

E2   For each vertex y in V2, there is at least one path $v = v_0, v_1, \ldots, v_k, y$ such that for $i = 0, 1, \ldots, k$, $v_i \in$ V1. E2 contains the edge $\overline{v_k y}$ from a shortest path of this form.

V3   The remaining vertices

E3   The remaining edges

Figure 3.15 gives examples of these sets. Whether or not G is a digraph it is helpful to think of the edges in E1 and E2 as having an orientation; the tail is the vertex closer to v. Edges in E2 lead from a vertex in V1 to one in V2. Edges in E1 and E2 will always be written to reflect this orientation; in other words, if we write $\overline{xy}$, we are assuming x is closer to v than y is.

Given the situation in Fig. 3.15(c), the next step is to select from V2 the vertex y that is closer to v than any other vertex not already in V1. We do not simply choose the edge in E2 with minimal weight as we did when finding a minimal spanning tree. (The incorrectness of such a choice is illustrated in Fig. 3.16, where repeatedly choosing a minimal weighted edge would give a path of length 10 from v to w, although there is a path of length 8.) We choose y by finding the edge $\overline{zy}$ that minimizes $d(v, z) + W(\overline{zy})$ over all edges $\overline{zy}$ in E2. This quantity is the weight of the path obtained by adjoining $\overline{zy}$ to the known shortest path to z.

Since the quantity $d(v, z) + W(\overline{zy})$ for $\overline{zy}$ in E2 may be used repeatedly, it can be computed once and saved. To compute it efficiently when $\overline{zy}$ is first put in E2, we also save $d(v, z)$ for z in V1. Thus we define

$$D(y) = d(v, y) \qquad \text{for } y \text{ in V1, and}$$
$$D(y) = d(v, z) + W(\overline{zy}) \qquad \text{for } y \text{ in V2,}$$

where $\overline{zy}$ is the edge to y in E2. The values of D are stored in an array.

After a vertex and the corresponding edge are moved from V2 to V1 and E2 to E1, some work must be done to reestablish the conditions that define V2 and E2. Figure 3.15(d) shows the sets after the vertex I and edge $\overline{GI}$ were selected. E2 contains the edge $\overline{AF}$, which gave the shortest path to F from vertices that were in V1. Now F may be reached from V1 via both A and I. $\overline{AF}$ must be replaced by $\overline{IF}$ in E2 because $\overline{IF}$ yields a shorter path to F. $D(F)$ must also be corrected. The vertex E which was in V3 must be moved to V2 because it is adjacent to I, now in V1. The edge $\overline{IE}$ is put in E2. Values of D for vertices added to V2 must be computed. Figure 3.15(e) shows the sets after these adjustments are made.

Does this method work? The questionable step is the selection of the next vertex and edge from V2 and E2 to be put in V1 and E1. For an arbitrary y in V2, $d(v, z) +$

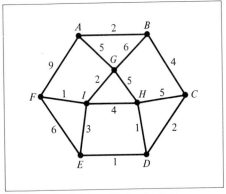

(a) The graph. Problem: Find a shortest path from A to H.

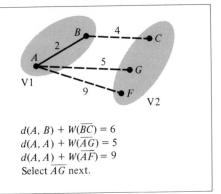

$d(A, B) + W(\overline{BC}) = 6$
$d(A, A) + W(\overline{AG}) = 5$
$d(A, A) + W(\overline{AF}) = 9$
Select $\overline{AG}$ next.

(b) An intermediate step.

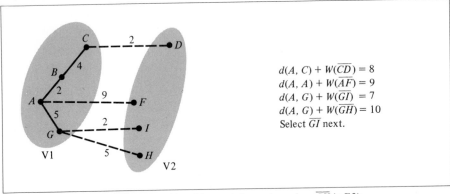

$d(A, C) + W(\overline{CD}) = 8$
$d(A, A) + W(\overline{AF}) = 9$
$d(A, G) + W(\overline{GI}) = 7$
$d(A, G) + W(\overline{GH}) = 10$
Select $\overline{GI}$ next.

(c) An intermediate step ($\overline{CH}$ was considered but not chosen to replace $\overline{GH}$ in E2).

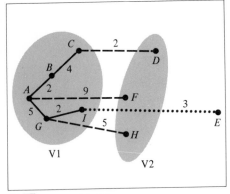

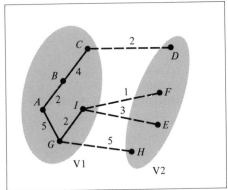

(d) $\overline{GI}$ was moved to E1.

(e) $\overline{AF}$ was replaced by $\overline{IF}$ in E2.

Figure 3.15
An example for the shortest-path algorithm.

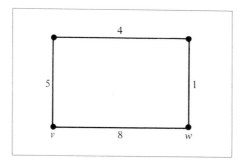

Figure 3.16

$W(\overline{zy})$ (where $\overline{zy}$ is the unique edge in E2 incident with y) is not necessarily equal to $d(v, y)$ because shortest paths to y might not pass through z. (In Figure 3.15, for example, the shortest path to H does not go through G, although $\overline{GH}$ is in E2 in Figs. 3.15(c), (d), and (e).) We claim that if y is chosen by minimizing $d(v, z) + W(\overline{zy})$ over all $\overline{zy}$ in E2, then the selection of $\overline{zy}$ does give a shortest path. This claim is proved in the following theorem.

Theorem 3.2 Let $G = (V, E, W)$ be a weighted graph or digraph with weights in $\mathbf{Z}^+$. Let V1 be a subset of V and let v be a member of V1. If $\widetilde{z}\widetilde{y}$ is chosen to minimize $d(v, z) + W(\overline{zy})$ over all z in V1 and y adjacent to z, then a shortest path from v to $\widetilde{y}$ can be obtained by adjoining $\widetilde{z}\widetilde{y}$ at the end of a shortest path from v to $\widetilde{z}$.

Proof. Look at Fig. 3.17. Suppose $\widetilde{y}$ and $\widetilde{z}$ are chosen as indicated and $v, x_1, \ldots, x_r, \widetilde{z}$ is a shortest path from v to $\widetilde{z}$. Let $P = v, x_1, \ldots, x_r, \widetilde{z}, \widetilde{y}$. $W(P) = d(v, \widetilde{z}) + W(\overline{zy})$.

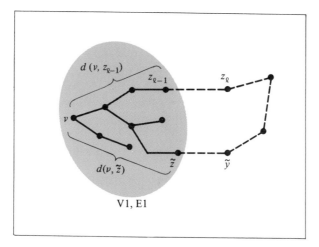

Figure 3.17
For the proof of Theorem 3.2.

Let $v, z_1, \ldots, z_\varrho, \ldots, \tilde{y}$ be any path from v to $\tilde{y}$; call it P'. Let z_ϱ be the first vertex in P' that is not in V1. (z_ϱ may be $\tilde{y}$. If $\varrho = 1$, interpret z_0 as v. In the algorithm, $z_{\varrho-1} z_\varrho$ would be in E2.) We must show that $W(P) \leqslant W(P')$.

$W(P) = d(v, \tilde{z}) + W(\tilde{z}\tilde{y}) \leqslant d(v, z_{\varrho-1}) + W(\overline{z_{\varrho-1} z_\varrho})$ (by the choice of $\tilde{z}$ and $\tilde{y}$) $\leqslant$
$W(P')$ since $v, z_1, \ldots, z_\varrho$ is part of the path P'. $\qquad\square$

The algorithm presented here uses virtually the same data structure as the minimal spanning tree algorithm; see Fig. 3.13. The only change is that the distance array D replaces W. In the shortest path algorithm E1 is a tree with v as the root and w a leaf. There is no way to tell which of the edges in E1 are in the path to w until the algorithm terminates, so all of the paths that branch out from v are retained by using the PARENT array as in the minimal spanning tree algorithm.

Algorithm 3.3 SHORTEST PATH

Input: $G = (V, E, W)$, a weighted graph or digraph; $v, w \in V$. G is represented by an adjacency list structure.

Output: A shortest path from v to w.

1.   [Initialization]
      $\text{VSET}(v) \leftarrow 1$; $x \leftarrow v$; $D(v) \leftarrow 0$; $\text{V2LINK}(0) \leftarrow \Lambda$
      for y in $V - \{v\}$; $\text{VSET}(y) \leftarrow 3$
2.   **while** $x \neq w$ **do**
3.   [Traverse the adjacency list for x.]
            $\text{PTR} \leftarrow \text{ADJLIST}(x)$
            while $\text{PTR} \neq \Lambda$ **do**
                  $y \leftarrow \text{VTX(PTR)}$
                  if $\text{VSET}(y) = 2$ and $D(x) + \text{WGT(PTR)} < D(y)$
                  then do [Replace the edge to y in E2 by $\overline{xy}$.]
                              $\text{PARENT}(y) \leftarrow x$; $D(y) \leftarrow D(x) + \text{WGT(PTR)}$
                        end
                  if $\text{VSET}(y) = 3$
                  then do [Put y in V2 and $\overline{xy}$ in E2.]
                              $\text{VSET}(y) \leftarrow 2$
                              $\text{V2LINK}(y) \leftarrow \text{V2LINK}(0)$; $\text{V2LINK}(0) \leftarrow y$
                              $\text{PARENT}(y) \leftarrow x$; $D(y) \leftarrow D(x) + \text{WGT(PTR)}$
                        end
                  $\text{PTR} \leftarrow \text{LINK(PTR)}$
            end
4.   **if** $\text{V2LINK}(0) = \Lambda$ **then return** 'no path to w'
5.   Find an edge $e = \overline{zy}$ in E2 such that $D(y)$ is minimal.
            Remove it.
            $\text{VSET}(y) \leftarrow 1$; $x \leftarrow y$
      end
6.   [Output the path. The following loop outputs the edges of the path in reverse order, i.e., from w to v.]

$[x = w$ at this point]
while $x \neq v$
  do output (PARENT$(x), x$); $x \leftarrow$ PARENT(x) **end**

Like Algorithm 3.2, Algorithm 3.3 runs in $\Theta(n^2)$ time in the worst case.

Although the shortest-path and minimal spanning tree algorithms are very similar in form, they behave differently; see Exercise 3.14 for an illustration.

3.4   *Traversing Graphs and Digraphs*

DEPTH-FIRST AND BREADTH-FIRST SEARCH

Almost any algorithm for solving a problem on a graph or digraph requires examining or processing each vertex or edge. In the two algorithms considered so far the order in which vertices and edges were considered was a fundamental part of the method used to solve the problem. Certain other strategies, or orders, for processing vertices and edges provide particularly useful and efficient methods for solving problems. Breadth-first search and depth-first search are two such traversal strategies.

In a depth-first search starting from the vertex v (which may be determined by the problem or chosen arbitrarily), a path is followed as far as possible, "visiting" or processing all the vertices along the way, until a "dead end" is reached. A "dead end" is a vertex such that all vertices adjacent to it have already been visited. At a dead end we back up along the last edge of the path and branch out in another direction. A depth-first traversal is illustrated in Fig. 3.18. In a breadth-first search vertices are visited in order of increasing distance from v, where distance is simply the number of edges in a shortest path. The central step of the breadth-first search, beginning with $d = 0$ and repeated until no new vertices are found, is to consider in turn each vertex, x, of distance d from v and, by traversing all edges incident with x, find and process all vertices of distance $d + 1$ from v. A breadth-first traversal is also illustrated in Fig. 3.18.

These descriptions of the two traversal methods contain some ambiguity. For example, if there are two vertices adjacent to v, which will be visited first? The answer will depend on implementation details; for example, the way in which the vertices are numbered or arranged in the representation of G. An efficient implementation for either method must keep a list of vertices that have been visited but whose adjacent vertices have not yet all been visited. Note that when a depth-first search backs up from a dead end it is supposed to branch out from the most recently visited vertex before pursuing paths from vertices that were visited earlier. Thus the list of vertices from which some paths remain to be traversed must be a stack. On the other hand, in a breadth-first search, in order to ensure that vertices close to v are visited before those farther away, the list must be a queue. Algorithms for both methods are presented below. In both, vertices are marked (using a one-bit mark field, or whatever is convenient) when first examined to prevent repeated work. The instruction "visit x" is used to indicate when the desired processing of the vertex x is done. Note that since a depth-first search does not back up from a vertex x until every edge from x has been traversed, it has a very simple recursive description:

DFS(v)

    Visit and mark v

    while there is an unmarked vertex w adjacent to v

        do DFS(w) **end**

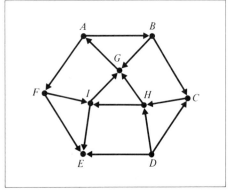

(a) A digraph.

Edges are numbered in the order traversed.

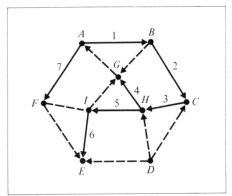

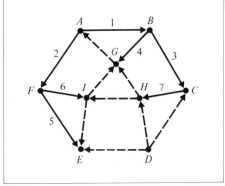

(b) Depth-first search beginning at A; order in which vertices are visited: $A\ B\ C\ H\ G\ I\ E\ F$

(c) Breadth-first search beginning at A; order in which vertices are visited: $A\ B\ F\ C\ G\ E\ I\ H$

Figure 3.18
Depth-first search and breadth-first search.

Algorithm 3.4 DEPTH-FIRST SEARCH

Input: $G = (V, E)$, a graph or digraph; $v \in V$, the vertex from which the search begins.

Notation: S is a stack, initially empty. The vertex on the top of the stack at any time is referred to by *top*.

1.   Visit, mark, and stack v.

2.   **while** S is nonempty **do**

3.   **while** there is an unmarked vertex w adjacent to *top*
4.   **do** visit, mark, and stack w **end** [w is now *top*.]
5.   Pop S
    end

Algorithm 3.5 BREADTH-FIRST SEARCH

Input: $G = (V, E)$, a graph or digraph; $v \in V$, the vertex from which the search begins.

Notation: Q is a queue, initially empty. "$x \Leftarrow Q$" means remove the front item from Q and denote it by x.

1.   Visit and mark v. Insert v in Q.
2.   **while** Q is nonempty **do**
3. $x \Leftarrow Q$
4.   **for** each unmarked vertex w adjacent to x **do**
5.   visit and mark w
6.   insert w in Q
      end
    end

Many variations and extensions may be made to these algorithms, depending on what they are used for. It is often necessary, for example, to do some sort of processing on each edge (or perhaps only on each edge that leads to an unmarked vertex) when it is traversed. The descriptions of the algorithms make no explicit mention of edges, but of course the implementation of the lines that require finding an unmarked vertex adjacent to a given vertex, say x, would involve examining edges incident with x and the necessary processing of edges would be done there. The algorithms given will visit only those vertices that can be reached by a path from v. If it is necessary to visit every vertex in G, instructions must be added to find an unmarked vertex after the traversal from v is complete and to begin again from the new starting point. All vertices must be unmarked before the first traversal, but the marks would not be erased before each succeeding one. In Fig. 3.18, for example, if a traversal were started at A all vertices except D would be visited. A second traversal beginning at D would have very little work to do if all the other vertices remain marked. Details for these extensions and implementation of depth-first search will be discussed later in an application to the problem of finding connected components of graphs. Implementation and analysis of breadth-first search are similar and are left as exercises. We emphasize depth-first search because its use leads to some very efficient and elegant algorithms, two of which are studied in Sections 3.5 and 3.6.

DEPTH-FIRST SEARCH AND RECURSION

We have seen that depth-first search can be simply described by a recursive algorithm. In fact there is a fundamental connection between recursion and depth-first search. In a recursive program the call structure can be diagrammed as a rooted tree where each vertex represents a recursive call to the program. The order in which the calls are executed corresponds to a depth-first traversal of the tree. Consider, for example, the recursive definition of the Fibonacci numbers: $F_0 = 0$, $F_1 = 1$, and for $n \geqslant 2$, $F_n = F_{n-1} + F_{n-2}$. The call structure for a recursive computation of F_6 is shown in Fig. 3.19. Each vertex is labeled with the current value of n, i.e., the n for which the subtree rooted at that vertex computes F_n. The order of the execution of the recursive calls is indicated in the figure. This ordering of the vertices in a tree is also known as *preorder*. (The reader should be aware that it is extremely inefficient to compute the Fibonacci numbers recursively; this example is used only to illustrate the connection between depth-first traversal and recursion.)

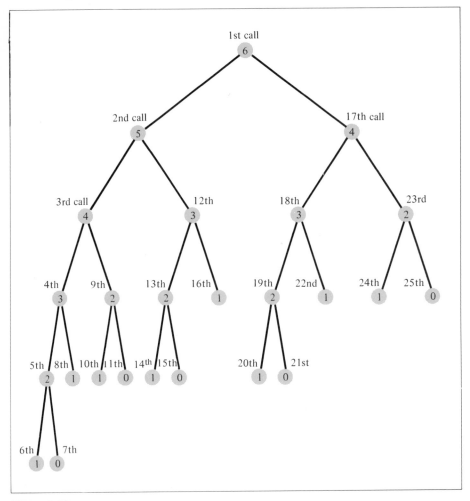

Figure 3.19
Call structure for recursive computation of Fibonacci numbers. Nodes are labeled by the index of the Fibonacci number to be computed.

Thus the logical structure of the solutions to a number of interesting problems solved by recursive routines is a depth-first search of a tree. The tree is not explicitly part of the problem nor is it explicitly represented as a data structure. Its relation to the solution of a problem is similar to the relation of a flowchart (a digraph) to a program; it gives an insight into what is going on.

Consider the problem of placing eight queens on a chessboard so that none is under attack by any other; in other words, so that none can reach another by moving along one row, column, or diagonal. It is not obvious that this can be done; we try as follows: Place a queen in the first (leftmost) square of the first (topmost) row. Then continue to place queens in each successive vacant row in the first (i.e., leftmost) column that is not under attack by any queen already on the board. Do this until all eight queens are on the board or all of the squares in the next vacant row are under attack. If the latter case occurs (which it does in ·the sixth row), go back to the previous row, move the queen there as few places as possible farther to the right so that it is still not under attack, and then proceed as before.

What tree is involved in this problem, and in what sense are we doing a depth-first search of it? The tree is shown in Fig. 3.20. Each vertex (other than the root) is labeled by a position on the chessboard. For $1 \leq i \leq 8$, the vertices at level i are labeled with board positions in row i. The children of a vertex, v, at level i are all board positions in row $i + 1$ which would not be under attack if there were queens in all board positions along the path from the root to v; in other words, the children are all the safe squares in the next row. In terms of the tree, the problem is to find a path from the root to a leaf of length eight. As an exercise the reader may write a recursive program for the queens problem such that the order in which the recursive calls are executed corresponds to a depth-first search. If there actually is a solution, only part of the tree in Fig. 3.20 is traversed. (Depth-first search, when used in a problem like this one, is also called back-track search.)

IMPLEMENTATION AND ANALYSIS OF DEPTH-FIRST SEARCH: FINDING CONNECTED COMPONENTS OF A GRAPH

Let $G = (V, E)$ be a graph (not directed) with $n = |V|$ and $m = |E|$. A *connected component* of G is a maximal connected subgraph, i.e., a connected subgraph that is not contained in any larger connected subgraph. The graph in Fig. 3.21, for example, has three connected components. The problem of finding the connected components of a graph may be solved by using depth-first search with very little embellishment. We may start with an arbitrary vertex, do a depth-first search to find all other vertices (and edges) in the same component, and then if there are some vertices remaining, choose one and repeat.

Various parts of the algorithm could require a lot of work if we choose a poor implementation. The innermost loop of the traversal algorithm (Algorithm 3.4) is controlled by the condition "**while** there is an unmarked vertex w adjacent to *top*." How can we determine efficiently if there is such a w? Certainly, we should use linked adjacency lists to represent the graph so that we can traverse the list for *top* looking for

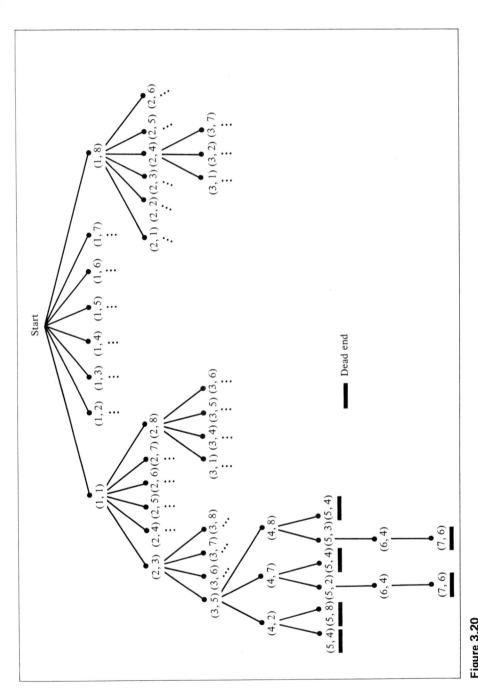

Figure 3.20
Tree for the eight-queens problem. Nodes are labeled by the coordinates of a board position in which a queen may be placed.

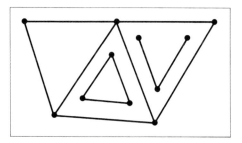

Figure 3.21
A graph with three connected components.

an unmarked w, but the quoted **while** statement may be executed several times for the same value of *top* (that is, each time the search backs up to *top* to branch out again). Do we have to traverse the adjacency lists (even partway) several times? Fortunately, no. Marked vertices are never unmarked, so the search for an unmarked w adjacent to *top* can begin where the previous search through the adjacency list for *top* ended. Thus, throughout the algorithm, the adjacency lists are traversed only once. An array of pointers, one for each vertex, is used to keep track of our place in each adjacency list.

The problem of finding an unmarked vertex from which to start a new depth-first search can be handled similarly; in other words, instead of checking through the list of vertices from the beginning each time a depth-first search is completed, we start wherever we left off the previous time. If the partition of the graph into connected components is to be recorded in the data structure for later use, it can be done by marking each vertex and/or edge with the number of the component to which it belongs or by linking or making a separate list of the vertices and/or edges in each component. The particular method chosen would depend on how the information is to be used later.

The depth-first search algorithm (Algorithm 3.4) is repeated below with the implementation made more explicit. It is preceded by the controlling program which uses it to number the vertices and produce a list of the edges in each connected component.

Algorithm 3.6 CONNECTED COMPONENTS

Input: $G = (V, E)$, a graph represented by the adjacency list structure of Fig. 3.8. V is assumed to be $\{1, 2, \ldots, n\}$.

Output: Lists of edges in each connected component. Also each vertex is numbered to indicate which component it is in.

Comments: A MARK array (n entries) is used to number the vertices. PTR is an array (with n entries) such that $\text{PTR}(v)$ points to the next node in the adjacency list for v from which the search will proceed the next time it tries to branch out from v. DEPTH-FIRST SEARCH uses a stack that is initially empty. The top vertex on the stack at any time is denoted by *top*.

1. **for** $v \leftarrow 1$ **to** n **do**
      [Initialize MARK and PTR arrays]
2.   MARK$(v) \leftarrow 0$; PTR$(v) \leftarrow$ ADJLIST(v)
   end
3. $j \leftarrow 1$ [j is used to number the vertices in a component.]
4. **for** $v \leftarrow 1$ **to** n **do**
5.   **if** MARK$(v) = 0$ **then do output** heading for jth component
6.   DEPTH-FIRST SEARCH(v, j)
7. $j \leftarrow j + 1$
                   end

    end

DEPTH-FIRST SEARCH (v, j)
1.   MARK$(v) \leftarrow j$; Stack v
2.   **while** stack is not empty **do**
3.   **while** PTR$(top) \neq \Lambda$ **do**
4. $w \leftarrow$ VTX(PTR(top))
5.   **output** (top, w) [output the edge]
6.   PTR$(top) \leftarrow$ LINK(PTR(top))
7.   **if** MARK$(w) = 0$ **then do** MARK$(w) \leftarrow j$
                            Stack w
                     end
          end
9.   Pop the stack
       end

    The number of operations done by CONNECTED COMPONENTS, excluding line 6, "DEPTH-FIRST SEARCH(v, j)," is clearly linear in n. In DEPTH-FIRST SEARCH (v, j), the number of instructions executed is proportional to the number of links traversed, since the instruction PTR$(top) \leftarrow$ LINK(PTR(top)) is executed once each time through the inner loop. Since for each vertex v, PTR(v) is initialized to ADJLIST(v) and never set back, the adjacency lists are traversed once and the complexity of the depth-first search, and hence the connected component algorithm, is $\Theta(\max\{n, m\})$, which is usually $\Theta(m)$.

    Since the data structure is essentially the adjacency list structure of Fig. 3.8, the amount of space used is $\Theta(n + m)$.

    Observe that the output of the depth-first search algorithm described will contain two copies of each edge because every edge is encountered twice. (See Fig. 3.8.) Here this may be a minor annoyance, but in some problems it is critical that an edge not be traversed twice. The reader should be able to modify the algorithm (and, if necessary, the data structure) to avoid the duplication in the output list without changing the order of the complexity of the algorithm (Exercise 3.19).

DEPTH-FIRST SEARCH TREES

The edges that lead to new, i.e., unmarked, vertices during a depth-first search of a graph or digraph G form a rooted tree called a depth-first search tree. If not all of the vertices can be reached from the starting vertex (the root) then a complete traversal of G partitions the vertices into several trees. For an undirected graph the search provides an orientation for each of its edges; they are oriented in the direction in which they are traversed. (If G is directed, its edges may be traversed only in the direction of their preassigned orientation.)

We say that a vertex v is an *ancestor* of a vertex w in a tree if v is on the path from the root to w; v is a proper ancestor of w if $v \neq w$. If v is a (proper) ancestor of w, then w is a (proper) descendant of v.

An edge of G that is directed from a vertex to one of its ancestors in a depth-first search tree is called a *back edge*. If G is not a directed graph, each of its edges will be a tree edge or a back edge. If G is a digraph, depth-first search partitions its edges into several classes: tree edges, back edges, edges that go from a vertex to one of its descendants other than a child, and edges, called *cross edges,* between two vertices such that neither is a descendant of the other. See Fig. 3.22 for illustrations. Note that

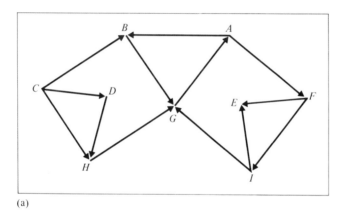

(a)

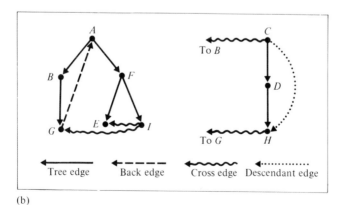

(b)

Figure 3.22
(a) A digraph. (b) Depth-first search trees for the digraph.

the head and tail of a cross edge may be in two different trees. The reader should prove that there can be no cross edges or descendant edges if G is not a digraph (Exercise 3.21). The distinctions between the various types of edges are important in some applications of depth-first search — in particular, in the algorithms studied in the next two sections.

3.5   *Biconnected Components of a Graph*

In the previous section we used depth-first search to partition a graph into connected components. Depth-first search may be used to partition graphs and digraphs in other useful ways also. We will study algorithms for such partitions in this section and in Section 3.6. A connected graph is said to be biconnected if the removal of any one vertex (and the edges incident with it) leaves a connected subgraph. Biconnectivity may be very important in a graph representing a communication or transportation network. Suppose, for example, that the vertices are telephone switching stations and the edges are telephone lines. If the graph is biconnected the system can still operate in the event that one station malfunctions. Hence the problem of determining if a graph is biconnected is important, as is finding those vertices, if any, that can disconnect it. Consideration of this example should suggest a dual problem to the reader: how to determine if there is an edge whose removal would disconnect the graph, and how to find such an edge if there is one. For example, if a railroad track is damaged, can trains still travel between any pair of stations? Relationships between the two problems are examined in Exercise 3.23.

Formally, a vertex v is an *articulation point* (also called a *cutpoint*) for a graph if there are distinct vertices w and x (distinct from v also) such that v is in every path from w to x. Clearly, the removal of an articulation point would leave an unconnected graph, so a connected graph is *biconnected* if and only if it has no articulation points. A *biconnected component* of a graph is a maximal biconnected subgraph, that is, a biconnected subgraph not contained in any larger biconnected subgraph. Alternatively, biconnected components may be defined in terms of an equivalence relation on the edges. Two edges e and e' are equivalent if $e = e'$ or if there is a cycle containing both e and e'. Then each subgraph consisting of the edges in one equivalence class and the incident vertices is a biconnected component. (Verifying that the relation described is indeed an equivalence relation and verifying that the two definitions of biconnected components agree are left as Exercises 3.25 and 3.26.) Figure 3.23 gives an illustration of biconnected components. Observe that although these components partition the edges, they do not partition the vertices; some vertices are in more than one component. (Which vertices are these?)

The backbone of the algorithm we will study for finding biconnected components is a depth-first search. We assume that the graph will be represented by an adjacency list structure as in the previous section. Enough information will be computed and

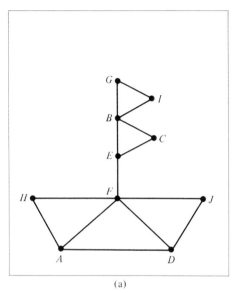 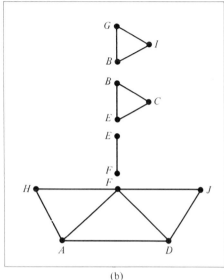

(a)   (b)

Figure 3.23
(a) A graph. (b) Its biconnected components.

saved so that the edges (and, implicitly, the incident vertices) can be divided into biconnected components as the search progresses. What information must be saved? How is it used to determine the biconnected components? Several wrong answers to these questions seem reasonable until they are examined carefully. We know that two edges are in the same component if they are in a cycle, and every cycle must include at least one back edge. Attempts to characterize biconnected components are considered in Exercise 3.27. The reader should work on that exercise before proceeding; it requires looking at a number of examples to determine relationships between back edges and biconnected components.

From now on we will use the shorter term "bicomponent" in place of "biconnected component."

THE BICOMPONENT ALGORITHM

A depth-first search passes through each vertex several times: when the vertex is first encountered and becomes part of the depth-first search tree, and several more times when the search backs up *to* it and attempts to branch out in a different direction. After the last of these encounters the search backs up *from* the vertex and does not pass through it or any of its descendants again. Thus processing, or computation, may be done when a vertex is first visited, when the search backs up to it, and/or when the search backs up from it. The bicomponent algorithm tests to see if a vertex in the tree is an articulation point each time the search backs up to it. Suppose the search is backing up to *v* from *w*. If there is no back edge from any vertex in the subtree rooted

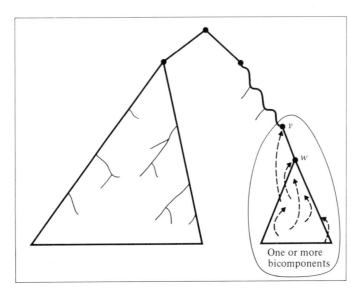

Figure 3.24
An articulation point in a depth-first search tree: Every path
from the root to w passes through v.

at w to a proper ancestor of v, then v must be on every path in G from the root to w
and is therefore an articulation point. See Fig. 3.24 for illustration. (The careful reader
should note that this argument is not valid if v is the root.) The subtree rooted at w
along with all back edges leading from it and along with the edge $\overline{vw}$ can be separated
from the rest of the graph at v, but it is not necessarily one bicomponent; it may be a
union of several. We ensure that bicomponents are properly separated by removing
each one as soon as it is detected. Since vertices at the outer extremities of the tree are
tested for articulation points before vertices closer to the root, this ensures that when
an articulation point is found, the subtree in question (along with the additional edges
mentioned above) forms one bicomponent.

This discussion implies that the algorithm must keep track of how far back in the
tree one can get from each vertex by following tree edges (directed away from the
root) and certain back edges. This information will be stored in an array BACK. The
vertices will be numbered in the order in which they are first visited. Values of BACK
will be these vertex numbers. For a vertex v, values of BACK(v) may be set when the
search is going forward and a back edge from v is encountered (as in Fig. 3.25b with
$v = F$ and in Fig. 3.25c with $v = C$) and when the search backs up to v (as in Fig.
3.25d with $v = B$), since any vertex that can be reached from a child of v can also be
reached from v. Determining which of two vertices is farther back in the tree is easy: If
v is a proper ancestor of w, the number assigned to v is less than the number assigned
to w. The numbers are stored in an array NUMBER. Thus we can tentatively formulate
the following rules for setting BACK(v):

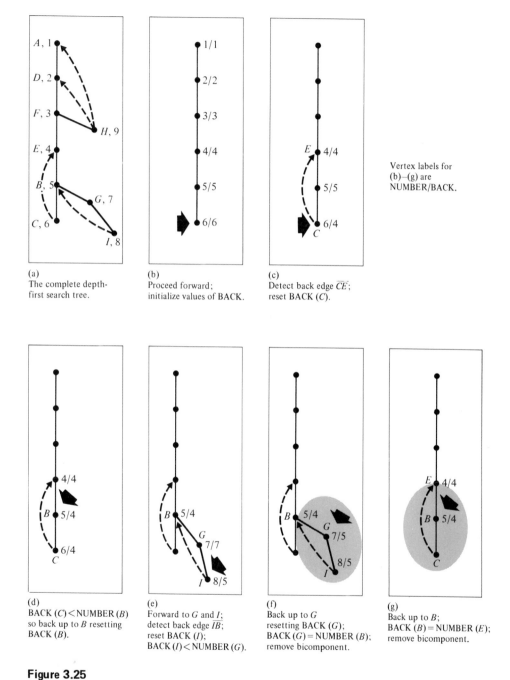

(a)
The complete depth-first search tree.

(b)
Proceed forward; initialize values of BACK.

(c)
Detect back edge $\overline{CE}$; reset BACK (C).

Vertex labels for (b)–(g) are NUMBER/BACK.

(d)
BACK (C) < NUMBER (B) so back up to B resetting BACK (B).

(e)
Forward to G and I; detect back edge $\overline{IB}$; reset BACK (I); BACK (I) < NUMBER (G).

(f)
Back up to G resetting BACK (G); BACK (G) = NUMBER (B); remove bicomponent.

(g)
Back up to B; BACK (B) = NUMBER (E); remove bicomponent.

Figure 3.25
The action of the bicomponent algorithm on the graph in Fig. 3.23. ➡ points to the vertex on the top of the stack.

1. When a back edge $\overrightarrow{vw}$ from v is detected, BACK(v) ← min{BACK(v), NUMBER(w)}.

2. When backing up from w to v, BACK(v) ← min{BACK(v), BACK(w)}.

(These rules imply that values of BACK must be properly initialized. BACK(v) will initially be set to NUMBER(v), but see Exercise 3.30.) Now, with BACK and NUMBER as described, the condition tested to detect a bicomponent when backing up from w to v is: BACK(w) ≥ NUMBER(v). (This condition is tested but not satisfied in Figs. 3.25d and 3.25e; it is satisfied in Figs. 3.25f and 3.25g.) When the test is satisfied, v is an articulation point (except perhaps if v is the root of the tree) and a complete bicomponent has been found and may be removed from further consideration. When this occurs, rule (2) above for resetting BACK(v) may be skipped.

The problem of exactly when and how to test for bicomponents is subtle but critical to the correctness of an algorithm. (See Exercises 3.32, 3.33, and 3.34.) The essence of the argument for the correctness of the method used by the algorithm presented here is contained in the following theorem.

Theorem 3.3 In a depth-first search tree, a vertex v, other than the root, is an articulation point if and only if v is not a leaf and some subtree of v has no back edge incident with a proper ancestor of v.

Proof. Suppose that v is an articulation point. Then there are vertices x and y such that v is on every path from x to y. Either x or y must be a proper descendant of v since otherwise there would be a path between them using (undirected) edges in the tree without going through v. Thus v is not a leaf. Let x be the descendant of v. If there were a back edge from the subtree of v containing x to a proper ancestor of v, there would be a path from x to y that does not pass through v; it uses tree edges, backward or forward, from x to the tail of the back edge, then the back edge, then again tree edges, backward or forward to y. (See Fig. 3.26 for some examples.) Thus, since v is an articulation point, the subtree containing x must satisfy the requirements of the theorem.

The remaining half of the proof is left as an exercise.   □

Theorem 3.3 does not tell us under what conditions the root is an articulation point. See Exercise 3.29.

The algorithm must keep track of the edges traversed during the search so that those in one bicomponent can easily be identified and removed from further consideration at the appropriate time. As the example in Fig. 3.25 illustrates, when a bicomponent is detected, its edges are the edges most recently processed. Thus edges are stacked as they are encountered, and when a bicomponent is detected when backing up to a vertex v, all of the edges from the top down to (but not including) the first one incident with a vertex closer to the root than v (i.e., with number less than NUMBER(v)), are in that bicomponent, and may be removed from the stack.

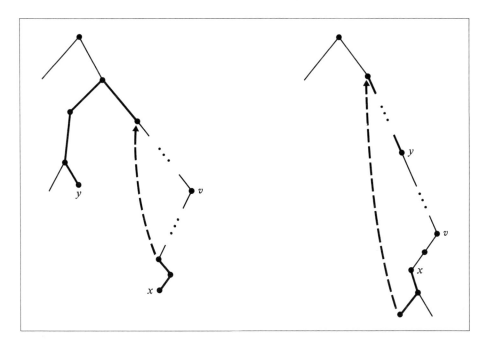

Figure 3.26
Examples for the proof of Theorem 3.3.

Algorithm 3.7 BICONNECTED COMPONENTS

Input: $G = (V, E)$, a connected graph represented by linked adjacency lists.

Output: Lists of the edges in each biconnected component of G.

Comment: S_V is a vertex stack (for controlling the depth-first search) and S_E is an edge stack; both are initially empty. The top vertex on S_V is denoted by *top*. Arrays NUMBER and BACK are used as described above.

1. Let x be an arbitrary vertex. NUMBER(x) ← 1; NUM ← 2; stack x on S_V.
    for v in $V - \{x\}$; NUMBER(v) ← 0.
2. FORWARD: **while** there is an unprocessed edge $\overline{top\ w}$ incident with *top* **do**
3.   Stack $\overline{top\ w}$ on S_E
4.   **if** NUMBER(w) > 0
5.   **then** BACK(*top*) ← min{BACK(*top*), NUMBER(w)}
6.   **else do** NUMBER(w) ← NUM; NUM ← NUM + 1
7.   BACK(w) ← NUMBER(w)
8.   Stack w on S_V
                 end
         end

9.   BACKUP: **if** there is more than one vertex on S_V
10.   **then do** $v \leftarrow$ second (from the top) vertex on S_V
11.   **if** BACK(top) $\geqslant$ NUMBER(v)
12.   **then do** output a heading for a new bicomponent,
13.   **while** both vertices incident with the top edge on
                           S_E have number $\geqslant$ NUMBER(v)
14.   **do** output and pop the top edge on S_E **end**
                      end
15.   **else** BACK(v) $\leftarrow$ min{BACK(v), BACK(top)}
16.   Pop S_V [v is now *top*.]
17.   **goto** FORWARD
                 end

As in Algorithm 3.6, line 2 can be implemented so that each adjacency list is scanned exactly once, but every edge of G is in two adjacency lists and it is necessary to avoid traversing and stacking an edge the second time that it is encountered. Failure to do so can result in incorrect output – some edges may be put in two different bicomponents. (See Exercise 3.31.) There are a few easy ways to handle this problem. One is to include links in the adjacency list nodes between the two entries for the same edge so that the second one may be removed from the data structure or marked to indicate that it should be ignored later.

ANALYSIS

As usual, let $n = |V|$ and $m = |E|$. The initialization, line 1, includes $\Theta(n)$ operations. The body of the **while** loop in the FORWARD section of the search is executed once for each edge in the graph, and the amount of work done within the loop (lines 3–8) is bounded by a constant. So the total time spent in the FORWARD section is $\Theta(m)$.

The label BACKUP (line 9) is not needed in the algorithm but is used to emphasize its structure. Control goes to the BACKUP section (lines 9–17) when the condition in the **while** statement in line 2 fails to hold, i.e., when all edges incident with the vertex on the top of the vertex stack have been traversed. This occurs once for each vertex since the stack is popped when the search backs up and numbered vertices are never stacked a second time. There is a loop within the BACKUP section in lines 13 and 14 which may not be executed each time the search backs up, and which requires varying amounts of time when it is executed. This loop does a fixed amount of work for each edge popped from the edge stack S_E. Since each edge is put on the stack exactly once (line 3) and S_E is empty when the algorithm terminates (i.e., when the test in line 9 fails), the total amount of work done in lines 13 and 14 is $\Theta(m)$. (The fact that S_E is empty when the test in line 9 fails is not obvious. The reader may check it by examining the behavior of the algorithm when control arrives at line 9 with two vertices on S_V.) Thus the complexity of Algorithm 3.7 is $\Theta(\max\{n, m\})$.

The amount of space used is $\Theta(n + m)$.

GENERALIZATIONS

The prefix "bi" means "two." Informally speaking, a biconnected graph has two disjoint paths between any pair of vertices (see Exercise 3.24). We can define triconnectivity (and, in general, k-connectivity) to denote the property of having three (in general, k) paths between any pair of vertices. An efficient algorithm that also uses depth-first search to find the triconnected components of a graph has been developed (see the notes and references at the end of the chapter), but it is much more complicated than the algorithm for bicomponents.

3.6   *Strongly Connected Components of a Digraph*

In the previous two sections we studied connectivity and biconnectivity of undirected graphs. We pointed out in Section 3.1 that the concept of connectivity can be applied to directed graphs in either of two ways, leading to the definitions of weakly connected and strongly connected. We will be concerned with the latter in this section. A digraph $G = (V, E)$ is strongly connected if and only if for any vertices v and w in V there is a path from v to w (and hence by interchanging the roles of v and w in the definition, there is a path from w to v as well). A *strongly connected component* (hereinafter called a *strong component*) of a digraph is a maximal strongly connected subgraph. We may give an alternative definition in terms of an equivalence relation, S, on the vertices. For v and w in V, let vSw if and only if there is a path from v to w and a path from w to v. Then a strong component consists of one equivalence class, $\tilde{V}$, along with all edges $\overline{vw}$ such that v and w are in $\tilde{V}$. See the example in Fig. 3.27. We will sometimes

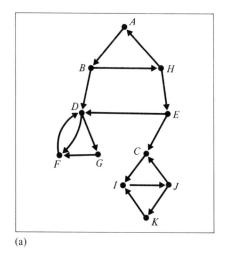

(a)

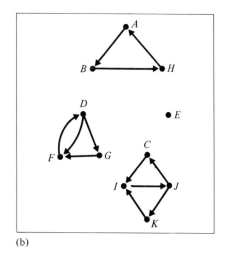

(b)

Figure 3.27
(a) A digraph. (b) Its strong components.

use the term strong component to refer only to the vertex set $\tilde{V}$; the meaning should be clear from the context.

The strong components of a digraph can each be collapsed to a point yielding a new digraph that has no cycles. Let $S_1, S_2, \ldots, S_p$ be the strong components of G. The *condensation* of G is the digraph $G' = (V', E')$, where V' has p elements, $s_1, \ldots, s_p$ and $\overline{s_i s_j}$ is in E' if and only if there is an edge in E from some vertex in S_i to some vertex in S_j. See Fig. 3.28 for an example. Solutions to some problems on digraphs may be simplified by treating the strong components and the condensation separately, taking advantage of the special properties of each (in other words, that the former are strongly connected and the latter acyclic). (Consider the relationship of the strong components and condensation of a program flowchart to the loop structure of the program. Also, see Section 6.1.)

In this section we study an algorithm for finding strong components that is very similar in structure to the biconnected component algorithm given in Section 3.5. The backbone of the algorithm here also is a depth-first search.

In an undirected graph the set of vertices in a depth-first search tree is exactly the set of vertices in one connected component of the graph and is independent of the particular starting point chosen for the search. Also, as we observed earlier, every edge in that component will be a tree edge or a back edge. With a digraph, where each edge may be traversed in only one direction, the situation is more complex. The partition induced on the vertex set by forming depth-first search trees for a digraph is not invariant; it depends on the starting points chosen for each search. (Consider the tree that would result from starting a depth-first search of the graph in Fig. 3.27 at E.) However, it is true that the set of vertices in any depth-first search tree is a union of strong components, so we may choose an arbitrary starting point, do a depth-first search, and by doing some computation similar to that done in the bicomponent algorithm, divide the tree into strong components.

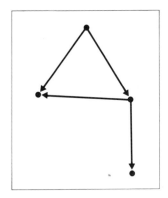

Figure 3.28
The condensation of the digraph in Fig. 3.27.

Let $T = (V_T, E_T)$ be a depth-first search tree for the digraph $G = (V, E)$. Recall that aside from tree edges and back edges, G may have cross edges and edges that lead from a vertex in V_T to one of its descendants. (See the end of Section 3.4 and Fig. 3.22.)

Assuming that branches are drawn (traversed) in left to right order, all cross edges point in the same direction — to the left. This fact can be stated formally if we define the relation $\mathscr{L}$ (read "is to the left of") on V_T as follows: For v and w in V_T, $v \mathscr{L} w$ if and only if w is encountered after v in the search and w is not a descendant of v. The following lemma is used in the arguments below that justify the algorithm. Its proof is easy and is left as an exercise.

Lemma 3.4 If v and w are in V_T and $\overline{vw} \in E$, then $w \mathscr{L} v$ or either v or w is a descendant of the other. (The only possibility excluded is that $v \mathscr{L} w$.)

For v in V_T let OLDEST(v) be the oldest (closest to the root) ancestor of v in T that can be reached by following tree edges, back edges, and cross edges. The strong components in T can be characterized as follows.

Theorem 3.5 For v and w in V_T, v and w are in the same strong component if and only if OLDEST(v) = OLDEST(w).

Proof. See Exercise 3.40.   □

Thus an algorithm could be devised that does depth-first searches of a digraph and computes the values of OLDEST. However, it is possible to determine the strong components with less information; OLDEST need not be explicitly computed for each vertex. The algorithm numbers vertices when they are first visited (storing the numbers in an array NUMBER) and uses an array LOW such that for a vertex v in T, LOW(v) is the lowest numbered vertex (not necessarily an ancestor of v) known to be in the same strong component as v. LOW(v) is initialized to the number of v when v is first encountered. When the algorithm backs up from v, it need only determine if it has just completed traversing a strong component. It does this by testing whether or not LOW(v) = NUMBER(v), for if after completely searching the subtree rooted at v no way has been found to reach a proper ancestor of v, there can be none. (Consider $v = D$ in Fig. 3.29. This argument uses the fact that cross edges go only to the left, i.e., to parts of T already examined. There can be no cross edge from the subtree leading to the right to nodes not yet visited.) Whenever the algorithm backs up from a vertex v such that LOW(v) = NUMBER(v), it removes all of the vertices in the subtree rooted at v from T. Each such set of vertices (along with the appropriate edges) is one strong component. (Can you prove this statement?)

When the algorithm backs up from a vertex v such that LOW(v) < NUMBER(v), the value of LOW(v) is carried back to the parent of v, as when it backs up from J to I in Fig. 3.29. Values of LOW are, of course, also updated when the search is proceeding forward. Suppose that when proceeding forward from a vertex v an edge leading to a numbered vertex w is encountered. If w is a descendant of v, then LOW(v) $\leq$ NUMBER(w) so LOW(v) need not be changed. If $\overline{vw}$ is a back edge, LOW(v) must be

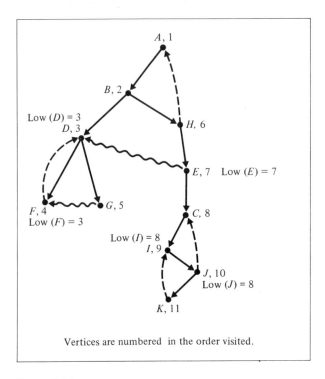

Vertices are numbered in the order visited.

Figure 3.29

set to min{LOW(v), NUMBER(w)}. If $\overline{vw}$ is a cross edge, things get a bit tricky. It is sometimes necessary and sometimes incorrect to set LOW(v) to min{LOW(v), NUMBER(w)}. For example, in Fig. 3.29, when the cross edge $\overline{GF}$ is encountered, LOW(G) must be reset to NUMBER(F) so that G will not be incorrectly considered as a strong component by itself. On the other hand, if LOW(E) is reset to min{LOW(E), NUMBER(D)} when the cross edge $\overline{ED}$ is encountered, E would incorrectly be put in the same strong component as A, B, and H. The following theorem provides the criterion for resetting LOW(v) (i.e., for setting LOW(v) to min{LOW(v), NUMBER(w)}; its value may not actually change).

Theorem 3.6 Suppose that when a cross edge $\overline{vw}$ is encountered during a depth-first search the algorithm sets LOW(v) ← min{LOW(v), NUMBER(w)} if and only if w has not already been assigned to another strong component. Then the case when LOW(v) is reset is exactly the case when v and w are in the same strong component and the algorithm works correctly.

Proof. Certainly if the strong component containing w has already been detected and removed, v is not in it. (This claim assumes that the algorithm has worked correctly so far; it can be formalized by using induction.) Suppose that w has not yet been

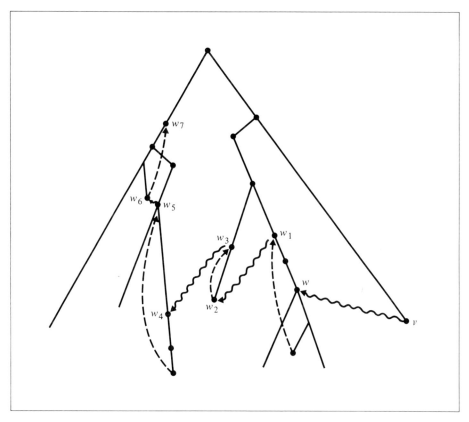

Figure 3.30
Illustration for the proof of Theorem 3.6. $w_k = w_7$; $w_i = w_3$.

removed from the tree. We will show that there is a path from w to v. Let $w_1 = w$, and let w_{i+1} = the vertex whose number is LOW(w_i), for each i until for some k, LOW(w_k) = NUMBER(w_k). (See Fig. 3.30.) By the definition of LOW, there is path from w to w_k. Either $w_k \mathscr{L} v$ or w_k is an ancestor of v. We will show that the latter must be true by assuming that $w_k \mathscr{L} v$ and deriving a contradiction. Consider the sequence $v = w_0$, $w = w_1, w_2, \ldots, w_k$. Let i be the largest index such that w_k is not an ancestor of w_i. Such an i exists since $w_k \mathscr{L} v$; i could be 0. (In Fig. 3.30, $i = 3$.). Then $w_k \mathscr{L} w_i$, and the search backed up from w_k before encountering w_i. Since LOW(w_k) = NUMBER(w_k), w_k and all of its descendants, including w_{i+1}, were removed from the tree. But then LOW(w_i) would not have been set to NUMBER(w_{i+1}). This contradicts the definition of w_{i+1}; thus, w_k must be an ancestor of v. Then there is a path, using tree edges, from w_k to v. Adjoining this path to the one from w to w_k (which may use cross edges, back edges, and tree edges) gives a path from w to v. Hence w and v are in the same strong component and setting LOW(v) ← min{LOW(v), NUMBER(w)} is correct.   □

The algorithm keeps track of the vertices to be assigned to the next strong component by recording each vertex in a list *SC* as it backs up from that vertex. When the root of a strong component is detected, *SC* contains all the vertices in that component (except the root) and may be output and emptied. Vertices will be tagged when they are assigned to a strong component so that cross edges may be handled properly.

Algorithm 3.8 STRONGLY CONNECTED COMPONENTS

Input: $G = (V, E)$, a digraph represented by linked adjacency lists.

Output: Lists of vertices in each strong component.

Comment: LOW and NUMBER are arrays used as described above. S is the stack used to control the depth-first search; it is initially empty. The top vertex on S will be denoted by *top*. *SC* is a list, initially empty, in which vertices in one strong component are stored until all vertices in that component have been found.

1. **for** each v in V; NUMBER(v) ← TAG(v) ← 0
   NUM ← 1
2. **while** there is a vertex x with number 0 **do**
3.   NUMBER(x) ← NUM; NUM ← NUM + 1; LOW(x) ← NUMBER(x)
4.   Stack x *[R stack]*
5.   FORWARD: **while** there is an unprocessed edge $\overline{top\ w}$ incident with *top* **do**   *[gets new w]*
6.   **if** NUMBER(w) = 0 [w is not yet in the tree.]
7.   **then do** NUMBER(w) ← NUM; NUM ← NUM + 1
8.   LOW(w) ← NUMBER(w)
9.   Stack w
           end   *[i.e., is on VSTK]*
10.   **else if** TAG(w) = 0 [w has not yet been output.]
11.   **then** LOW(*top*) ← min{LOW(*top*), NUMBER(w)}
         end   *[No unprocessed edge]*
12.   BACKUP: **if** LOW(*top*) = NUMBER(*top*) [found a strong component]
13.   **then do** TAG(*top*) ← 1; **output** *top*   *[how??]*
14.   **for** each vertex x in *SC*
15.   **do** TAG(x) ← 1; **output** x **end**
16.   set *SC* empty
           end
17.   **else do** put *top* in *SC*
18. z ← second (from the top) vertex on S
19.   LOW(z) ← min{LOW(z), LOW(*top*)}
           end
20.   Pop S [backs up from *top* to z, the new *top*]
21.   **if** S is not empty **then goto** FORWARD
       end

IMPLEMENTATION, TIMING, AND SPACE USAGE

Much of the discussion of implementation, timing, and space usage of the bicomponent algorithm carries over with small changes to the strong component algorithm. Control for the outer loop, line 2, can be implemented as in the connected component algorithm (Algorithm 3.6, line 4) so that each vertex is examined only once. Thus we conclude that the number of operations done by the strong component algorithm is $\Theta(\max\{n, m\})$. Extra space, aside from the adjacency list structure, is used for S, SC, TAG, NUMBER, LOW, and PTR. (PTR is not mentioned explicitly here, but used as in the connected component algorithm to control the traversals of the adjacency lists at line 5.) NUMBER, LOW, and PTR have n entries each. Since no vertex is on S and SC at the same time (except *top*, briefly; i.e., from line 17 to line 20), S and SC can share an array with n cells. TAG needs only one bit per vertex, so we conclude that roughly $4n$ extra cells are used by Algorithm 3.8.

3.7   *Exercises*

Section 3.1: Definitions and Representations

3.1   Indicate which of the graphs and digraphs in the figures in Section 3.1 are connected, weakly connected, or strongly connected.

3.2   Euler paths
a) A popular game among grade school children is to draw the following figure
without picking up one's pencil and without retracing a line. Try it.

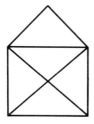

b) The diagram in Fig. 3.31 provides a similar but slightly harder problem. The
diagram shows a river with two islands in it connected to each other and to
the banks by seven bridges. The problem is to determine if there is a way to
start out on either bank of the river or on either island and take a walk so
that you cross each bridge exactly once. (No swimming allowed.) Try it.
c) The problems in (a) and (b) may be studied abstractly by examining the
graphs below. G_2 is obtained by representing each bank and island as a vertex
and each bridge as an edge. (Some pairs of vertices are connected by two
edges, but this departure from the definition of a graph will not cause trouble
here.) The general problem is: Given a graph (with multiple edges between
pairs of vertices permitted), find a path through the graph that traverses each
edge exactly once. Such a path is called a Euler path. (The term "path" is
used in a broader sense here than in the text; it may pass through a vertex more

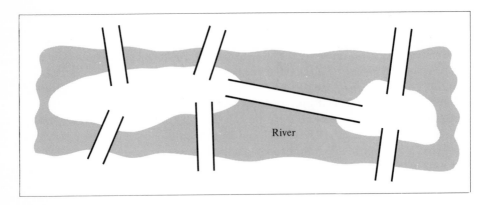

Figure 3.31
The Konigsberg bridges (for Exercise 3.2).

than once.) This problem is solvable for G_1 but not for G_2; that is, there is no way to walk across each bridge exactly once. Find a necessary and sufficient condition for a graph to have an Euler path.

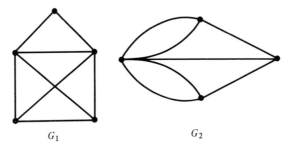

G_1 G_2

3.3. Suppose that a digraph G represents a binary relation R. Describe a condition on G that holds if and only if R is transitive.

Section 3.2: A Minimal Spanning Tree Algorithm

("Minimal spanning tree" is abbreviated "MST" in the exercises.)

3.4. Describe a sequence of connected weighted graphs G_n such that G_n has n vertices and the running time of the MST algorithm (Algorithm 3.2) for these graphs is linear in n.

3.5. Describe a sequence of connected weighted graphs G_n such that G_n has n vertices and the MST algorithm does no comparisons of weights when G_n is the input. (The algorithm will require time at least proportional to n because it must succeed in finding a minimal spanning tree.)

3.6. How many comparisons would be done at line 6 of the MST algorithm on the graph $G = (V, E, W)$ where $V = \{v_1, v_2, \ldots, v_n\}$, $E = \{\overline{v_1 v_i}: i = 2, \ldots, n\}$, and for $i = 2, \ldots, n$, $W(\overline{v_1 v_i}) = 1$? (Working through this problem may suggest to the reader that saving information about the ordering of the weights of edges in E2 could decrease the work done at line 6. See Exercise 3.7.)

3.7. How many comparisons would be done at line 6 of the MST algorithm on the complete graph with vertices $v_1, \ldots, v_n$ where for $1 \leqslant i < j \leqslant n$, $W(\overline{v_i v_j}) = n + 1 - i$?

3.8. Complete the proof of Theorem 3.1 by showing that T' is a spanning tree.

3.9. Prove or disprove: The MST algorithm will work correctly even if weights may be negative.

Section 3.3: A Shortest-Path Algorithm

3.10. Will the shortest-path algorithm (Algorithm 3.3) work correctly if weights may be negative? Justify your answer by an argument or a counterexample.

3.11. What path from P to Q would be found by the shortest-path algorithm (Algorithm 3.3) in Fig. 3.32? (Weights are written next to the edges.)

3.12. Consider the problem of finding just the distance, but not a shortest path, from v to w in a weighted graph or digraph. Outline a modified version of Algorithm 3.3 to do this with the aim of eliminating as much work and extra space usage as possible. Indicate what changes, if any, you would make in the data structure used by Algorithm 3.3, and indicate what work or space you would eliminate.

3.13. Some graph algorithms are written with the assumption that the input is always a complete graph (where an edge has weight ∞ or 0 to indicate its absence from the graph for which the user really wants to solve the problem). Such algorithms are usually shorter and "cleaner" because there are fewer cases to consider. In the algorithms in Sections 3.2 and 3.3, for example, there would be no set V3 since all vertices would be adjacent to vertices in V1.
   a) With the aim of simplifying as much as possible, rewrite the shortest-path algorithm with the assumption that $G = (V, E, W)$ is a complete graph and W maps E into $\mathbf{Z}^+ \cup \{\infty\}$. Describe any changes you would make in the data structures used.
   b) Compare your algorithm and data structures with those in the text, using the criteria of simplicity, time (worst case and other cases), and space usage (for graphs with many edges that have weight ∞ and for graphs with few).

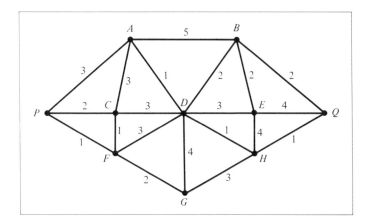

Figure 3.32

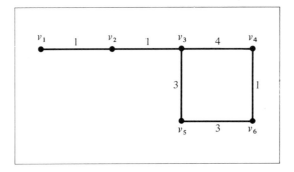

Figure 3.33

3.14. For the graph in Fig. 3.33 indicate which edges would be in E1 when Algorithm 3.2 terminates and which would be in E1 when Algorithm 3.3 terminates after finding a shortest path from v_1 to v_6.

Section 3.4: Traversing Graphs and Digraphs

3.15. Is is always, sometimes, or never true that the order in which vertices are put into V1 in the minimal spanning tree algorithm (Algorithm 3.2) corresponds to the order in which the vertices would be visited by breadth-first or depth-first search? Justify your answer with examples or a proof.

3.16. Do Exercise 3.15 for the shortest-path algorithm (Algorithm 3.3).

3.17. Give an example of a graph in which a depth-first search backs up from a vertex before all the vertices that can be reached from it via one or more edges are marked.

3.18. a) Choose and describe appropriate data structures for implementing a breadth-first search and rewrite Algorithm 3.5 using instructions that indicate the operations performed on the data structures. (Writing a documented program is one way to do this problem.)
  b) What is the order of the running time for your implementation?
  c) How would your algorithm (or program) be modified to produce as output a list of all edges examined in the order in which they are considered?

3.19. Write a depth-first search algorithm for an undirected graph such that the output is a list of the edges encountered, with each edge appearing once. Describe any changes in the adjacency list structure of Fig. 3.8 needed by your algorithm.

3.20. Write a depth-first search algorithm whose output is a list of the edges traversed that lead to unmarked vertices.

3.21. Prove that if G is a connected graph (not directed), each of its edges either is in the depth-first search tree or is a back edge.

3.22. Write an algorithm to determine whether a given graph $G = (V, E)$, with $n = |V|$ and $m = |E|$, is a tree. Would you use the same algorithm if you could assume that the graph is connected? If not, write one that uses that assumption also.

Section 3.5: Biconnected Components of a Graph

3.23. A connected graph is *edge-biconnected* if removal of any one edge leaves a connected subgraph. Which, if either, of the following statements is true? Give a proof or counterexample for each.
   a) A biconnected graph is edge-biconnected.
   b) An edge-biconnected graph is biconnected.

3.24. Is the following property on a graph $G = (V, E)$ necessary and sufficient for G to be biconnected? Prove your answer.

   For each pair of distinct vertices v and w in V, there are two distinct paths from v to w that have no vertices in common except v and w.

3.25. Show that for a graph $G = (V, E)$ the following relation, R, on E is an equivalence relation: $e\,R\,e'$ if and only if $e = e'$ or there is a cycle containing e and e'. How many equivalence classes are there in the graph below?

3.26. Show that a subgraph of a graph G consisting of the edges in one equivalence class of the relation described in Exercise 3.25 and the incident vertices is a maximal biconnected subgraph of G.

3.27. The following two definitions of functions on the vertices in a depth-first search tree of a graph may be used in an attempt to provide necessary and/or sufficient conditions for two vertices to be in the same biconnected component of the graph. Show by exhibiting counterexamples that $OLD_i(v) = OLD_i(w)$ is neither necessary nor sufficient for v and w to be in the same biconnected component $(i = 1, 2)$.
   a) $OLD_1(x) =$ the "oldest" – i.e., closest to the root – ancestor of x that can be reached by following tree edges (directed away from the root) and back edges, or x itself if no such path leads to an ancestor of x.
   b) $OLD_2(x) =$ the oldest ancestor of x that can be reached by following directed tree edges and *one* back edge, or x itself if no such path leads to an ancestor of x.

3.28. Complete the proof of Theorem 3.3.

3.29. a) Find a necessary and sufficient condition for the root of a depth-first search tree for a connected graph to be an articulation point. Prove it.
   b) Note that the bicomponent algorithm terminates at line 9 if only one vertex remains on the stack. Verify that the algorithm works properly when the root is an articulation point by describing how it behaves (i.e., which lines are executed) when it backs up to the root.

3.30. a) Would the bicomponent algorithm work properly if for each vertex v, BACK(v) were initialized to the number of the parent of v in the depth-first search tree? If so, explain why; if not, give an example in which it does not work.
   b) What if BACK(v) were initialized to ∞ for all v?

3.31. Give an example of a graph that shows that the bicomponent algorithm may produce incorrect answers if no attempt is made to avoid traversing and stacking an edge the second time it is encountered in the adjacency list structure.

3.32. The test to detect a bicomponent when backing up from w to v (line 11 of Algorithm 3.7, letting $w = top$) is BACK(w) $\geqslant$ NUMBER(v). Can the case BACK(w) $>$ NUMBER(v) ever occur when the algorithm is executed? If so, explain in what circumstances it occurs; if not, explain why not.

3.33. Would Algorithm 3.7 work properly if the test for a bicomponent, line 11, were changed to BACK(v) $\geqslant$ NUMBER(v)? If so, explain why; if not, give an example in which it does not work.

3.34. In line 5 of Algorithm 3.7 BACK(top) is set to min{BACK(top),NUMBER(w)}, where w is the head of a back edge. Would the algorithm work correctly if BACK(top) were set to min{BACK(top),BACK(w)} instead? If so, explain why; if not, give an example in which it will not work.

3.35. Write the bicomponent algorithm in recursive form.

3.36. See the definition of edge-biconnected in Exercise 3.23. Suppose we define an edge-bicomponent of a graph to be a maximal edge-biconnected subgraph. Devise an algorithm for finding the edge-bicomponents of a graph. (Note that each vertex is in exactly one edge-bicomponent, but some edges may not be in any.)

Section 3.6: Strongly Connected Components of a Digraph

3.37. Prove that the condensation of a digraph is acyclic.

3.38. Prove that the set of vertices in a depth-first search tree for a digraph is a union of strong components.

3.39. Prove Lemma 3.4.

3.40. Prove Theorem 3.5.

3.41. Write the strong component algorithm in recursive form.

3.42. We observed that if $\overline{vw}$ is in G and w is a descendant of v in a depth-first search tree, it does not matter whether or not LOW(v) is reset to min{LOW(v), NUMBER(w)} when the edge $\overline{vw}$ is encountered. What does the algorithm actually do? In other words, is line 11 of Algorithm 3.8 executed in such a case?

3.43. Draw a diagram showing the strong components of the digraph in Fig. 3.22(a).

3.44. Find the strong components of the digraph in Fig. 3.27 by carefully following the steps of the algorithm.

3.45. Extend or modify Algorithm 3.8 so that it outputs a list of all the edges, as well as the vertices, in each strong component. Try to minimize the amount of extra time used to do so.

Additional Problems

3.46. Formulate the following problem as a graph problem. Indicate clearly what graph you would use and what graph theoretic problem you would solve to obtain the desired answer.

A job consists of a series of tasks $S_1, S_2, \ldots, S_k$ where the amount of time required by task S_i is t_i, for $1 \leqslant i \leqslant k$. A list of pairs (S_i, S_j) is given such that if (S_i, S_j) is on the list, task S_i must be completed before task S_j is begun.

Pairs that would be implied by transitivity are *not* on the list. For example, if (S_i, S_j) and (S_j, S_p) are given, then (S_i, S_p) is not. Assume that there is an unlimited number of processors available so that several tasks can be done at the same time. How much time is needed to complete the job?

(It may be useful to include dummy tasks B (for "begin") and E (for "end"), each of which takes no time.)

3.47. We mentioned in Section 1 that if a graph or digraph is represented by an adjacency matrix, then almost any algorithm that operates on the graph will have worst-case complexity at least $\Theta(n^2)$, where n is the number of vertices. There are, however, some problems that can be solved quickly even when the adjacency matrix is used. Here is one.

a) Let $G = (V, E)$ be a digraph with n vertices. A vertex s is called a "sink" if for every v in V such that $s \neq v$, there is an edge $\overline{vs}$, and there are no edges of the form $\overline{sv}$. Give an algorithm to determine whether or not G has a sink, assuming that G is given by its $n \times n$ adjacency matrix.

b) Count the matrix entries examined by your algorithm in the worst case. It is easy to give an algorithm that looks at $\Theta(n^2)$ entries, but there is a linear solution.

3.48. Find the best lower bound you can for the number of adjacency matrix entries that must be examined to solve the problem described in Exercise 3.47(a). Prove that it is a lower bound.

3.49. Outline an efficient way of solving the following problem: Given a digraph G, determine whether or not G has a cycle. How much time does your method take?

3.50. a) Design an efficient algorithm to assign numbers to the vertices of an *acyclic* digraph so that $\#(v) < \#(w)$ if $v \neq w$ and there is a path from v to w. Describe any data structures you use other than the standard adjacency list structure.

b) Give the order of the worst-case complexity of your algorithm as a function of n or m (or both), where n is the number of vertices and m the number of edges.

3.51. Design an efficient algorithm to find a path in a connected graph that goes through each edge exactly once in each direction. (Here, "path" means a sequence of vertices, $v_1, \ldots, v_k$ where v_{i+1} is adjacent to v_i $(1 \leqslant i \leqslant k - 1)$; a vertex may appear in the path more than once.)

PROGRAMS

Each of the following program assignments requires a subprogram that reads in the number of vertices and a list of pairs representing the edges in a given graph (or digraph), along with weights if appropriate, and that sets up the adjacency list structure. The student should write this routine so that, with small changes, it could be used for any of the problems.

Test data should be chosen so that all aspects of a program are tested. Include some of the examples in the text.

1. The minimal spanning tree algorithm, Algorithm 3.2. Output should include the graph, the set of edges in the tree, along with their weights, and the total weight of the tree.

2. The shortest-path algorithm, Algorithm 3.3. Output should include the graph (or digraph), the vertices between which a shortest path is sought, the edges in the path, along with their weights, and the total weight of the path.

3. Breadth-first search (Exercise 3.18)

4. Depth-first search (Exercise 3.19)

5. The bicomponent algorithm, Algorithm 3.7

6. The strong component algorithm, Algorithm 3.8

7. Exercise 3.49

8. Exercise 3.50

NOTES AND REFERENCES

The shortest path and minimal spanning tree algorithms are from Dijkstra (1959), but that paper does not discuss implementation of the algorithms. In some applications it is necessary to find a spanning tree with minimal weight among those that satisfy other criteria required by the problem, so it is useful to have an algorithm that generates spanning trees in order by weight so that each can be tested for the other criteria. Gabow (1977) presents algorithms that do this.

The adjacency list structure used in this chapter was suggested by Tarjan and is described, along with the algorithms in Tarjan (1972) and Hopcroft and Tarjan (1973b). Hopcroft and Tarjan (1973a) presents an algorithm for finding the triconnected components of a graph. See Hopcroft and Tarjan (1974) for a very efficient algorithm to test graphs for planarity — another important problem for graphs.

Harary (1969) is a good text on graph theory. For a variety of applications and algorithms, see Even (1973), Aho, Hopcroft, and Ullman (1974), Deo (1974), Reingold, Nievergelt, and Deo (1977), and Ford and Fulkerson (1962).

String Matching

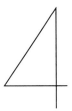

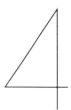

4.1   *The Problem and a Straightforward Solution*

In this chapter we study the problem of detecting the occurrence of a particular string of characters, called the pattern, in another string of characters, called the subject. The problem arises often in the processing of text of any kind; for example, in the text editors on interactive computer systems, in business-letter editing systems, etc. Let P be the pattern and S the subject, and let m be the length of (i.e., the number of characters in) P and n the length of S. We will assume that n is fairly large relative to m.

The reader should think about the problem and write out, or at least outline, an algorithm to solve it before proceeding. Your algorithm will probably be very similar to the first, fairly straightforward one we present here.

Starting at the beginning of each string, we compare characters, one after the other, until either the pattern is exhausted or a nonmatch is found. In the former case we are done; a copy of the pattern has been found in the subject. In the latter case we start over again, comparing the first pattern character with the second subject character. Whenever a nonmatch is found, we (figuratively) move the pattern one more place forward in the subject and start again comparing the first pattern character with the next subject character.

Example Comparisons are done (in left to right order) on the pairs of characters indicated by arrows.

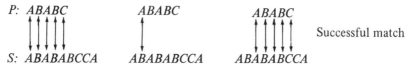

Observe that moving the pattern all the way past the point where the mismatch occurred could fail to detect an occurrence of the pattern.

Algorithm 4.1 STRAIGHTFORWARD STRING MATCHING ALGORITHM

Input: P and S, the pattern and subject strings, both non-null.

Output: A success or failure indicator. If the pattern is found, i, its location in the string, may be output also.

Comment: Characters in S and P are denoted by subscripted s's and p's, respectively. The index variables used are: i, the current guess at where in S P begins; and j and k, indexes for S and P, respectively. The algorithm refers to the lengths of the strings (n and m). We assume that either these are given as input or the ends of the strings are marked in some way and references to n and m can be replaced by tests for the ends of the strings.

1. $i \leftarrow 0$
2.   **while** $i < n$ **do**
3. $i \leftarrow i + 1$; $j \leftarrow i$; $k \leftarrow 1$
4.   **while** $s_j = p_k$ **do**
5.   **if** $k = m$ **then return** 'SUCCESS'
6.   **else do** $j \leftarrow j + 1$; $k \leftarrow k + 1$ **end**
     end
     end
7.   **return** 'FAILURE'

If the lengths of the strings are given, the test in line 2 could read "**while** $i < n - m + 1$" to terminate the algorithm when i gets too close to the end of S for an occurrence of P to be possible.

Observe that with a few modifications, Algorithm 4.1 can be used to detect a particular sublist in a linked list.

ANALYSIS

The basic operation to be counted is character comparison. The running time of the algorithm is bounded by a multiple of the number of comparisons since the test at line 4 is done once in every pass through the inner loop and at least once in every pass through the outer loop. If the pattern appears at the beginning of the subject, m comparisons are done. If p_1 is not in S at all, n comparisons are done. These are cases in which the algorithm works quickly. What is the worst case? The number of comparisons would be maximized if for each value of i – that is, each possible starting place for P in S – all but the last character of P matched the corresponding subject characters. Thus the number of character comparisons in the worst case is *at most mn*. To claim that the worst case requires *mn* comparisons, we must show that the situation described can really occur, i.e., that P and S can be constructed so that all but the last character of P matches corresponding characters beginning anywhere in S. For some algorithms, inputs that require a lot of work at one step may require very little work at another step. Thus adding up the maximum possible work at each step gives an upper bound but not necessarily an exact value for the work done in the worst case. We show that *mn* character comparisons really are done in the worst case by exhibiting strings that require that many: $P = {}'AA \ldots AB'$ ($m - 1$ A's followed by a B) and $S = {}'A \ldots A'$ (nA's).

Algorithm 4.1 has a property that in some applications is very undesirable: It may often be necessary to back up in the subject string (line 3: $j \leftarrow i$). Thus if the entire subject is not easily accessible, a segment of it as long as the pattern must be stored and continually updated. The algorithm we present in the next section was devised specifically to eliminate the need to back up in the subject. It turned out to be faster as well.

4.2   *The Knuth–Morris–Pratt Algorithm*

PATTERN MATCHING WITH FINITE AUTOMATA

We first describe briefly, without formal algorithms, an approach to the pattern matching problem that has some important good points and some drawbacks. The construction used by the main algorithm of this section was suggested by the method we describe now and salvages some of its advantages while eliminating the disadvantages.

Given a pattern *P*, it is possible to construct a finite automaton that can be used to scan the subject for a copy of *P* very quickly. A finite automaton can easily be interpreted as a special kind of machine or flowchart, and a knowledge of automata theory is not necessary to understand this method. Let Σ be the alphabet, or set of characters, from which the characters in *P* and *S* may be chosen and let $k = |\Sigma|$. The flowchart, or finite automaton, has three types of nodes:

START

* A STOP node, which means "STOP; a match was found."

A READ node, which means "Read the next subject character. If there are no further characters in the subject string, halt; there is no match."

The flowchart has *k* arrows leading out from each READ node; each is labeled with a character from Σ and it indicates which node to go to next if the character on the arrow is the character just read from the subject. The reader should study the example in Fig. 4.1 to understand why the arrows point where they do. The READ nodes serve as a sort of memory. For instance, if execution reaches the third READ node, the last two characters read from the subject were A's. What preceded them is irrelevant. For a successful match, they must be followed immediately by a B and a C. If the next character is a B we can move on to node 4, which remembers that *AAB* has appeared. On the other hand, if the next character read at node 3 were a C, we would have to return to node 1 and wait for another A to begin the pattern.

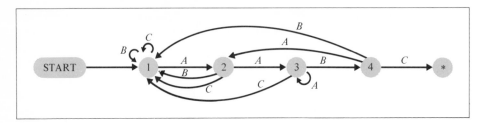

Figure 4.1
The finite automaton to detect occurrences of $P = $ '*AABC*'.

Once the flowchart for the pattern is constructed, the subject can be tested for an occurrence of the pattern by examining each subject character only once, hence in $O(n)$ time. This is a big improvement over Algorithm 4.1 both in timing and in the fact that once a subject character has been examined, it never has to be reconsidered; in other words, there is no backing up in the subject. The difficulties occur in trying to find efficient methods for constructing the finite automaton — that is, for deciding where all the arrows go — and for representing it in the computer so that the correct arrow to follow at each step may be found quickly, without having to search all k of them. The first of these is the more significant problem; there are well-known algorithms to construct the finite automaton to recognize a particular pattern, but in the worst case these algorithms require a lot of time. If P is known in advance or if m is very small compared to n (or if the same pattern will be sought in a number of different subject strings), then construction of the finite automaton may be worthwhile. In the general case in which P is to be used once and m is somewhat but not drastically smaller than n, the construction takes too much time.

Observe that the difficulties with the finite automaton approach arise from the fact that there is an arrow for each character in Σ leading out from each READ node. It takes time to determine where each arrow should point and it takes time to find the appropriate arrow to follow when scanning the subject. Also, space to represent mk arrows is needed. Thus a better algorithm will have to eliminate some of the arrows.

THE KNUTH–MORRIS–PRATT FLOWCHART

When constructing the finite automaton for a pattern P it is easy to put in the arrows that correspond to a successful match. For example, when drawing Fig. 4.1 for the pattern '*AABC*' the first step is to draw

$$\text{START} \longrightarrow \bullet \xrightarrow{A} \bullet \xrightarrow{A} \bullet \xrightarrow{B} \bullet \xrightarrow{C} \circledast$$

The difficult part is the insertion of the rest of the arrows. The Knuth-Morris-Pratt algorithm (which, for brevity, will be called the KMP algorithm) also constructs a sort of flowchart to be used to scan the subject. The KMP flowchart contains the easy arrows — i.e., the ones to follow if the desired character is read from the subject — but it contains only one other arrow from each node, an arrow to be followed if the desired character was *not* read from the subject. The arrows are called the success links and the failure links, respectively. The KMP flowchart differs from the finite automaton in several details: The character labels of the KMP flowchart are on the nodes rather than on the arrows; the next character from the subject is read only after a success link has been followed; the same subject character is reconsidered if a failure link is followed; there is an extra node which causes a new subject character to be read; and the starting point is the node that corresponds to the first pattern character. As in the finite automaton, if the ⊛ is reached, a copy of the pattern has been found; if the end of the subject is reached elsewhere in the flowchart, the scan terminates unsuccessfully. This informal description of the scanning procedure should enable the reader to use the KMP flowchart in Fig. 4.2 to scan a subject string. Try '*ACABAABABA*' and refer to Table 4.1 if you have difficulty.

We now need a computer representation of the KMP flowchart, an algorithm to construct it (to determine how to set the failure links), a formal algorithm for the scan procedure, and an analysis of the two algorithms.

CONSTRUCTION OF THE KMP FLOWCHART

The flowchart representation is quite simple; it uses two arrays, one containing the characters of the pattern and one containing the failure links. The success links are implicit in the ordering of the array entries.

Table 4.1

Action of the KMP flowchart in Fig. 4.2 for the pattern '*ABABCB*' on the subject '*ACABAABABA*'

KMP cell number	Subject character being scanned		Success (s) or failure (f)
	Index	Character	
1	1	*A*	s
2	2	*C*	f
1	2	*C*	f
0	2	*C*	get next char.
1	3	*A*	s
2	4	*B*	s
3	5	*A*	s
4	6	*A*	f
2	6	*A*	f
1	6	*A*	s
2	7	*B*	s
3	8	*A*	s
4	9	*B*	s
5	10	*A*	f
3	10	*A*	s
4	end of subject		failure

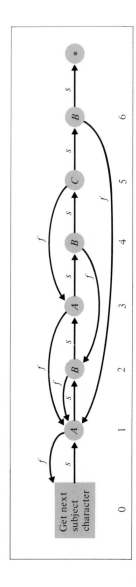

Figure 4.2
The KMP flowchart to detect occurrences of *P* = '*ABABCB*'.

Let FLINK be the array of failure links. FLINK(i) will be the index of the node pointed to by the failure link at the ith node, for $1 \leqslant i \leqslant m$. The special node which merely forces the next subject character to be read is considered to be the zero-th node; FLINK(1) = 0. To see how to set the other failure links we consider an example. Let $P =$ '*ABABABCB*' and suppose that the first six characters have matched six consecutive subject characters as indicated:

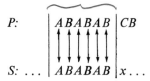

$$
\begin{array}{ll}
P: & | ABABAB\ |CB \\
 & \uparrow\uparrow\uparrow\uparrow\uparrow\uparrow \\
 & \downarrow\downarrow\downarrow\downarrow\downarrow\downarrow \\
S:\ \ldots & | ABABAB\ |x\ldots
\end{array}
$$

Suppose that the next subject character, x, is not a C. The next possible place where the pattern could begin in the subject is at the third position shown, that is, as follows:

$$
\begin{array}{ll}
P: & | ABAB\ |ABCB \\
 & \\
 & \\
S:\ \ldots AB & | ABAB\ |x\ldots
\end{array}
$$

The pattern is moved forward so that the longest initial segment which matches part of the subject preceding x is lined up with that part of the subject. Now x should be tested to see if it is an A — in particular, the third A of the pattern. Thus the failure link for the node containing the C should point to the node containing the third A. In general, the failure link of the ith node is set to the number j, which satisfies:

1. $j < i$

2. The first $j - 1$ characters in the pattern match the $j - 1$ characters that precede the ith node, i.e., $p_1, \ldots, p_{j-1}$ match $p_{i-(j-1)}, \ldots, p_{i-1}$.

3. j is the largest integer satisfying (1) and (2).

Note that $j = 1$ vacuously satisfies (2). Condition (3) provides for the maximum overlap of an initial segment of the pattern with the part of the subject just scanned. An occurrence of the pattern could be missed if j were chosen too small. (Consider what would happen if in the example above the failure link for C was set to point to the second A, and if $x = A$ and is followed by a C and B in the subject.)

A little thought should convince the reader that conditions (1), (2), and (3) correctly characterize the proper setting of the failure links, but the characterization does not suggest an algorithm for efficiently setting them. Suppose that the first $i - 1$ failure links have been set. By (2) [with $j = \text{FLINK}(i - 1)$] the following segments of p must match:

$$
\begin{array}{llll}
p_{i-\text{FLINK}(i-1)} & p_{i-\text{FLINK}(i-1)+1} & \cdots & p_{i-2} \\
\uparrow & \uparrow & & \uparrow \\
\downarrow & \downarrow & & \downarrow \\
p_1 & p_2 & \cdots\ p_{\text{FLINK}(i-1)-1}
\end{array}
$$

$$\tag{4.1}$$

If node i in the flowchart is reached when scanning a subject string, the last subject character read had to be p_{i-1}. If $p_{i-1} = p_{\text{FLINK}(i-1)}$, the two matching sequences in (4.1) can be extended by one more character to get:

$$
\begin{array}{ccccc}
p_{i-\text{FLINK}(i-1)} & \cdots & p_{i-2} & \bigg| & p_{i-1} \\
\big\uparrow & & \big\uparrow & & \big\uparrow \\
\big\downarrow & & \big\downarrow & & \big\downarrow \\
p_1 & \cdots & p_{\text{FLINK}(i-1)-1} & \bigg| & p_{\text{FLINK}(i-1)}
\end{array}
\qquad (4.2)
$$

and we have the following rule: If $p_{i-1} = p_{\text{FLINK}(i-1)}$, then set FLINK($i$) to FLINK($i-1$) + 1. In Fig. 4.3, FLINK(6) = 4 because p_1,p_2,p_3 matches p_3,p_4,p_5. Since $p_6 = p_4$, FLINK(7) was set to 5. What if $p_{i-1} \neq p_{\text{FLINK}(i-1)}$, as in Fig. 4.3, for $i = 8$? We must find an initial substring of p that matches a substring ending at p_{i-1}. The match in (4.1) can't be extended, so we look farther back. Let $j = \text{FLINK}(i-1)$ and let $j' = \text{FLINK}(j)$. By the properties of the failure links we have the following:

$$
\begin{array}{cccccc}
p_{i-j} \cdots & p_{i-j'} & \cdots p_{i-2} & \bigg| & p_{i-1} \\
\big\uparrow & \big\uparrow & \big\uparrow & & \\
\big\downarrow & \big\downarrow & \big\downarrow & & \\
p_1 \cdots & p_{j-j'+1} \cdots & p_{j-1} & \bigg| & \text{not } p_{i-1} \\
& \big\uparrow & \big\uparrow & & \\
& \big\downarrow & \big\downarrow & & \\
& p_1 & \cdots p_{j'-1} & \bigg| & ?
\end{array}
\qquad (4.3)
$$

Clearly if $p_{j'} = p_{i-1}$ we would have an initial substring to match a substring ending at p_{i-1} and FLINK(i) should be $j' + 1$. If $p_{j'} \neq p_{i-1}$ we must follow the failure link from node j' and try again. This process is continued until we find a failure link j such that $p_j = p_{i-1}$ or (as in Fig. 4.3 for $i = 8$) $j = 0$. In either case, FLINK(i) should be $j + 1$.

Algorithm 4.2 KMP "FLOWCHART" CONSTRUCTION

Input: P, a string of characters; $m \geq 1$, the length of P.

Output: FLINK, the array of failure links.

Comment: If m is not given, the test in line 2 can be replaced by a test for the end of P.

1.   FLINK(1) $\leftarrow$ 0; $i \leftarrow 2$
2.   while $i \leqslant m$ do
3. $j \leftarrow \text{FLINK}(i-1)$
4.   while $j \neq 0$ and $p_j \neq p_{i-1}$
5.   do $j \leftarrow \text{FLINK}(j)$ end
6.   FLINK(i) $\leftarrow j + 1$
7. $i \leftarrow i + 1$
      end

ANALYSIS OF THE FLOWCHART CONSTRUCTION

For the timing analysis of the flowchart construction we count character comparisons (line 4). Note that the comparison is done once in every pass through the inner loop

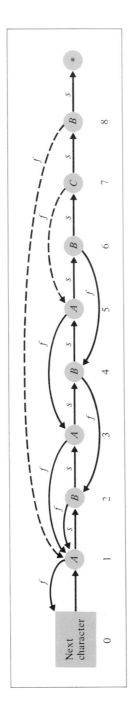

Figure 4.3
Setting failure links in a KMP flowchart.

and at least once in every pass through the outer loop, so the running time is bounded by a multiple of the number of comparisons.

We call a comparison "successful" if $p_j = p_{i-1}$ and "unsuccessful" otherwise. After each successful comparison i is incremented; thus $m - 1$ successful comparisons are done (for i ranging from 2 to m). After every unsuccessful comparison j is decreased [line 5, FLINK(j) $< j$ by condition (1) for failure links], so we can bound the number of unsuccessful comparisons by determining how many times j can decrease. Observe the following:

1. j is first set to 0 (the first time line 3 is executed).

2. j is increased in line 3 (after its first execution) by exactly 1 each time. The relevant steps are:

    line 6: FLINK(i) $\leftarrow j + 1$

    line 7: $i \leftarrow i + 1$

    line 3: $j \leftarrow$ FLINK($i - 1$),

    so line 3 sets j to $j + 1$.

3. Line 3 is in the outer loop and is executed $m - 1$ times.

4. j is never negative.

Since j starts at 0 and is increased by 1 $m - 1$ times and is never negative, j cannot be decreased more than $m - 1$ times. Thus the number of unsuccessful comparisons is at most $m - 1$ and the total is at most $2m - 2$. Observe that to count character comparisons, we actually counted the number of times that the indexes i and j changed. The latter is another good measure of the work done by the algorithm. The analysis showed that i and j change a total of at most $3(m - 1)$ times. The important point is that the complexity of the construction of the flowchart is linear in the length of the pattern.

THE KMP SCAN ALGORITHM

We have already informally described the procedure for using the KMP flowchart to scan the subject. The algorithm follows.

Algorithm 4.3 KMP SCAN ALGORITHM

Input: P and S, the pattern and subject strings; $m \geqslant 1$ and $n \geqslant 0$, their lengths; FLINK, the array of failure links set up in Algorithm 4.2.

Output: A success or failure indicator.

1. $i \leftarrow 1$; $j \leftarrow 1$ [j indexes flowchart nodes; i indexes subject characters]
2. **while** $i \leqslant n$ **do**
3.   **while** $j \neq 0$ and $p_j \neq s_i$
4.   **do** $j \leftarrow$ FLINK(j) **end**

5.   **if** $j = m$ **then return** 'SUCCESS'
6.   **else do** $i \leftarrow i + 1$; $j \leftarrow j + 1$ **end**
     end
7.   **return** 'FAILURE'

The analysis of the scan algorithm uses an argument very similar to that used for the flowchart construction, and it is left to the reader. The number of character comparisons done by Algorithm 4.3 is at most $2n$.

REMARKS AND EXTENSIONS

The Knuth-Morris-Pratt pattern matching algorithm, which is comprised of Algorithms 4.2 and 4.3, does $\Theta(n + m)$ operations in the worst case, a significant improvement over the $\Theta(mn)$ worst-case complexity of Algorithm 4.1, and the KMP algorithm never has to back up in the subject.

Two extensions to the pattern matching problem are often useful: Find *all* occurrences of the pattern in the subject, and find any one of a finite set of patterns in the subject. It is easy to modify Algorithm 4.3 so that it finds all nonoverlapping occurrences of the pattern. The second extension can be accomplished by constructing a more complex flowchart in the form of a tree. Such a flowchart for the set {*AACC*, *AAAB*, *ABCBA*, *ABAC*, *ACA*} is shown in Fig. 4.4. Designing an algorithm to construct the flowchart and one to use it to scan a subject is left as an exercise. Some points to consider: How will the tree be represented in the computer? (It is not a binary tree.) How will the end of a complete occurrence of a pattern be detected? (What if '*AA*' were one of the patterns in a set?) The running time of the construction algorithm should be proportional to the sum of the lengths of the patterns. The scan time should be linear in the length of the subject.

4.3   *Exercises*

Section 4.2: The Knuth-Morris-Pratt Algorithm

4.1.   Draw the finite automaton (flowchart) for the pattern *ABAABA*, where $\Sigma = \{A, B, C\}$.

4.2.   Prove that Algorithm 4.2 sets the failure links for the KMP flowchart so that conditions (1), (2), and (3) in the text are satisfied.

4.3.   Draw the KMP pattern matching flowchart for each of the following patterns:
   a) *AAAB*
   b) *AABAACAABABA*

4.4.   Show that the KMP scan algorithm (Algorithm 4.3) does at most $2n$ character comparisons.

4.5.   a) Exactly how many character comparisons are done by the KMP flowchart construction algorithm for the pattern $P =$ '$A \ldots AB$' ($m - 1$ *A*'s followed by one *B*)?
   b) Exactly how many character comparisons are done by the KMP scan algorithm to test the subject string $S =$ '$A \ldots A$' (n *A*'s) for an occurrence of P?

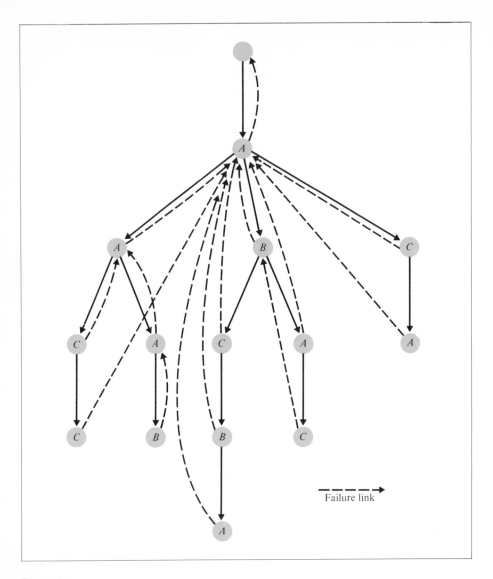

Failure link

Figure 4.4

    c) Given an arbitrarily large m_0, find a pattern Q with m letters for some $m \geqslant m_0$ such that the KMP flowchart construction algorithm does more character comparisons for Q than it does for the P with m letters described in (a).

4.6. Rewrite Algorithm 4.3 so that it finds all nonoverlapping occurrences of the pattern in the subject; (that is, all that do not overlap occurrences already detected; for example, in ABABABA, ABA should be found beginning at the first and fifth places).

4.7. a) Write an algorithm to construct a pattern matching flowchart like the KMP flowchart but for detecting any one of a finite set of patterns in a subject string. (See the last paragraph of Section 4.2.)

    b) Write a scan algorithm for the pattern matching flowchart in (a).

4.8. Rewrite the KMP algorithms to work on inputs that are simply linked lists. Assume node format ⟦ KEY │ LINK ⟧ and assume that S and P are pointers to the first node of the subject list and pattern list, respectively.

PROGRAMS

Write a program for Exercise 4.6. The output should include a copy of each pattern tested, along with the FLINK array for the pattern, a copy of each subject tested and the index in the subject of the first character of the pattern for each (nonoverlapping) occurrence of the pattern.

NOTES AND REFERENCES

The KMP algorithm presented here is from Morris and Pratt (1970). For a more accessible and more recent treatment of the algorithm, its history, a faster variation of it, and related problems, see Knuth, Morris, and Pratt (1977). Boyer and Moore (1977) contains another interesting string matching algorithm. Aho, Hopcroft, and Ullman (1974). Aho and Corasick (1975), and Galil (1976) contain variations of the KMP algorithm and related problems.

Polynomials and Matrices

5

5.1   *Introductory Comments*

The problems examined in this chapter are polynomial evaluation (with and without preprocessing the coefficients), polynomial multiplication (as an illustration of the discrete Fourier transform), and multiplication of matrices and vectors. The operations usually used for such tasks are multiplication and addition. Multiplication, on most computers, requires more time than addition does, and some of the algorithms presented "improve" upon the straightforward or most widely known methods by reducing the number of multiplications at the expense of some extra additions. Hence their value depends on the relative costs of the two operations. Other algorithms presented reduce the number of both operations (for large input sizes).

Many lower-bound results are stated without proof in this chapter. See the notes and references at the end of the chapter for further comment and references on these.

5.2   *Evaluating Polynomial Functions*

Let $p(x) = a_n x^n + a_{n-1} x^{n-1} + \cdots + a_1 x + a_0$ be a polynomial with real coefficients and $n \geqslant 1$. Suppose the coefficients $a_0, a_1, \ldots, a_n$ and x are given and the problem is to evaluate $p(x)$. It may seem that a reasonable measure of work is the number of multiplications and additions done, but some algorithms may use division and subtraction and do fewer multiplications and additions. Thus, particularly when discussing lower bounds, we will consider the total number of multiplications and divisions and the total number of additions and subtractions. The two types of operations will be denoted $*/$ and $\pm$, respectively.

Perhaps the most obvious way to solve the problem is to compute each term and add it to the sum of the others already computed. The following algorithm does this.

Algorithm 5.1 POLYNOMIAL EVALUATION – TERM BY TERM

Input: The coefficients of $p(x)$ in the array A; X; and $n \geqslant 1$.

Output: P, the value of $p(x)$.

$P \leftarrow A(0) + A(1) * X$; XPOWER $\leftarrow X$
for $I \leftarrow 2$ **to** N **do**
        XPOWER $\leftarrow$ XPOWER $* X$
        $P \leftarrow P + A(I) *$ XPOWER
end

Algorithm 5.1 does $2n - 1$ multiplications and n additions.

HORNER'S METHOD

Is there a better way? Is there a way to compute $ab + ac$, given a, b, and c, with less than two multiplications? Yes, of course; by factoring it as $a(b + c)$. Similarly, the key to Horner's method for evaluating $p(x)$ is simply a particular factorization of p:

$$p(x) = [\cdots ((a_n x + a_{n-1})x + a_{n-2})x + \cdots + a_1]x + a_0.$$

The computation is done in a short loop with only n multiplications and n additions.

Algorithm 5.2 POLYNOMIAL EVALUATION – HORNER'S METHOD

Input: A, X, and n, as in Algorithm 5.1.

Output: P.

$P \leftarrow A(N)$
for $I \leftarrow N - 1$ **to** 0 **by** -1; $P \leftarrow P * X + A(I)$

Thus just by factoring p we have cut the number of multiplications in half without increasing the number of additions. Can the number of multiplications be reduced further? Can the number of additions be reduced?

LOWER BOUNDS FOR POLYNOMIAL EVALUATION

Lower bounds on the number of required operations may be derived by examining formal evaluation schemes that may do multiplication, addition, and subtraction using x, the coefficients, some constants, and intermediate results computed from these. Division is not permitted here. Formally, a scheme for evaluating $p(x) = a_n x^n + a_{n-1} x^{n-1} + \cdots + a_1 x + a_0$ using constants $c_1, \ldots, c_t$ is a finite sequence of steps as follows:

$$s_i = a_i \qquad\qquad 0 \leqslant i \leqslant n$$
$$s_{n+1} = c_i \qquad\qquad 1 \leqslant i \leqslant n$$
$$s_{n+t+1} = x$$
$$s_i = s_j \circ s_k \qquad i > n+t+1,\ i > j, k, \text{ and } \circ \text{ is one of the}$$

three permitted operators.

The last step should compute $p(x)$. The following scheme evaluates $p(x) = a_2 x^2 + a_1 x + a_0$ by Horner's method.

$$s_0 = a_0$$
$$s_1 = a_1$$
$$s_2 = a_2$$
$$s_3 = x$$
$$s_4 = s_2 \cdot s_3$$
$$s_5 = s_4 + s_1$$
$$s_6 = s_5 \cdot s_3$$
$$s_7 = s_6 + s_0$$

We say that a step $s_i = s_j \circ s_k$ uses a_m if $j = m$ or $k = m$ (i.e., s_j or s_k is a_m) or if s_j or s_k uses a_m.

We will show that at least n $\pm$'s are necessary to evaluate a polynomial of degree n. It can be shown by a similar but more complicated argument that a scheme to evaluate a polynomial of degree n must do at least n multiplications. It is also known that if divisions are permitted, at least n $*$/'s are required. Thus Horner's method is optimal and, since division takes at least as much time as multiplication, it uses the best mix of these two operators.

Theorem 5.1 establishes the lower bound on $\pm$'s by proving a slightly stronger statement by induction.

Theorem 5.1 A scheme to evaluate $p(x) = a_n x^n + a_{n-1} x^{n-1} + \cdots + a_{n-r} x^{n-r}$, for $0 \leqslant r \leqslant n$, contains at least r $\pm$ steps.

Proof. For $r = 0$, the statement is clearly true. Suppose it is true for $r - 1$ where $r > 0$. Let S be a scheme to evaluate $p(x)$ for any values of x, a_n, $a_{n-1}, \ldots, a_{n-r}$. There must be a $\pm$ step that uses a_{n-r}; if not, $p(x)$ would have to be a multiple of a_{n-r} or independent of a_{n-r}, and it is not true that one of those cases always holds. Let $s_i = s_j \pm s_{n-r}$ (or $s_i = s_{n-r} \pm s_j$) be the first $\pm$ step that uses a_{n-r}. We will eliminate this step and make a few modifications to the scheme to obtain a scheme to evaluate $q(x) = a_n x^n + \cdots + a_{n-(r-1)} x^{n-(r-1)}$. Let $a_{n-r} = 0$ and include 0 among the constants used. Then if $s_i = s_j \pm s_{n-r}$ or $s_i = s_{n-r} + s_j$, the ith step may be eliminated and s_i may be replaced by s_j in all other steps. If $s_i = s_{n-r} - s_j$, this step may be replaced by a multiplication of s_j by -1 (which must be included among the available constants). Similar modifications are made for all other $\pm$ steps that have a_{n-r}, that is, s_{n-r}, as an operand. Any step of the form $s_i = s_{n-r} \cdot s_k$ may be changed to $s_i = s_z \cdot s_k$, where $s_z = 0$. The new scheme evaluates $q(x)$ and, by the induction assumption, has at least $r - 1 \pm$ steps. At least one $\pm$ was eliminated from S so S had at least $r \pm$ steps.   □

PREPROCESSING OF COEFFICIENTS

Preprocessing some of the data in a problem means, informally, that some of the input is known in advance and a specialized program may be written. Suppose that a problem has inputs I and I' and we denote an algorithm for the problem by A. When we speak of preprocessing I, we mean finding an algorithm A_I with input I' which produces the same output as A with inputs I and I'. Thus the preprocessing problem has two parts: the algorithm A_I which depends on I, and an algorithm that, with I as input, produces the algorithm A_I. A and A_I are, rigorously speaking, solving different problems and, as we shall see, their complexities may differ.

In some situations one polynomial has to be evaluated for a large number of different arguments. One example is a power series approximation of a function. In such cases, preprocessing of the coefficients may reduce the number of $*/$'s required for each evaluation.

Let $p(x) = a_n x^n + a_{n-1} x^{n-1} + \cdots + a_1 x + a_0$, where $n = 2^k - 1$ for some $k \geqslant 1$. Thus, p has 2^k terms, some of which may be zero. The procedure for evaluating $p(x)$ described here assumes that p is monic, i.e., that $a_n = 1$. Extending the algorithm to the general case is left as an exercise.

If $n = 1$, then $p(x) = x + a_0$ and is evaluated by doing one addition. Suppose $n > 1$ and p is written as follows for some j and b:

$$p(x) = (x^j + b) \cdot q(x) + r(x),$$

where q and r are monic polynomials of degree $2^{k-1} - 1$ (i.e., with half as many terms as p, counting zero terms). Then $p(x)$ may be evaluated by carrying out the following steps:

1.   Evaluate $q(x)$ and $r(x)$;

2.   Compute x^j;

3.   Multiply $(x^j + b)$ by $q(x)$ and add $r(x)$.

Since q and r satisfy the same conditions as p, i.e., they are monic and their degree is $2^{k'} - 1$ for some k', the same scheme could be used recursively to evaluate them. How must j and b be chosen to ensure that q and r have the desired properties? Clearly $j = \text{degree}(p) - \text{degree}(q) = 2^k - 1 - (2^{k-1} - 1) = 2^{k-1}$. Note that since j is a power of 2, x^j can be computed fairly quickly. The correct value for b becomes clear when we divide $p(x)$ by $x^j + b$ to obtain $q(x)$, the quotient, and $r(x)$, the remainder.

$$\overbrace{x^{2^{k-1}-1} + a_{2^k-2}x^{2^{k-1}-2} + \cdots + a_{2^k-1}}^{q(x)}$$

$$x^{2^{k-1}} + b \overline{\smash{\big)}\, x^{2^k-1} + a_{2^k-2}x^{2^k-2} + \cdots + a_{2^{k-1}}x^{2^{k-1}} \mid + a_{2^{k-1}-1}x^{2^{k-1}-1} \quad + \cdots + a_1x + a_0}$$

$$x^{2^k-1} + a_{2^k-2}x^{2^k-2} + \cdots + a_{2^{k-1}}x^{2^{k-1}} \mid + bx^{2^{k-1}-1} \qquad + \cdots + ba_{2^{k-1}}$$

$$0 \qquad\qquad \underbrace{(a_{2^{k-1}-1} - b)x^{2^{k-1}-1} + \cdots + (a_0 - ba_{2^{k-1}})}_{r(x)}$$

For r to be monic, $a_{2^{k-1}-1} - b$ must be 1, so $b = a_{2^{k-1}-1} - 1$. Thus the preprocessing algorithm factors p as follows:

$$p(x) = [x^{2^{k-1}} + (a_{2^{k-1}-1} - 1)] \cdot q(x) + r(x)$$

and factors q and r recursively by the same procedure. The factorization is complete when q and r have degree 1. The following example illustrates the entire procedure.

Example $p(x) = x^7 + 6x^6 + 5x^5 + 4x^4 + 3x^3 + 2x^2 + x + 1$

$k = 3$; $j = 2^{k-1} = 4$; $b = a_{2^{k-1}-1} - 1 = a_3 - 1 = 2$. Thus $x^j + b = x^4 + 2$.

$$\overbrace{x^3 + 6x^2 + 5x + 4}^{q(x)}$$

$$x^4 + 2\overline{\smash{\big)}\, x^7 + 6x^6 + 5x^5 + 4x^4 \mid + 3x^3 + 2x^2 + x + 1}$$

$$x^7 + 6x^6 + 5x^5 + 4x^4 \mid + 2x^3 + 12x^2 + 10x + 8$$

$$\underbrace{x^3 - 10x^2 - 9x - 7}_{r(x)}$$

Thus $p(x) = (x^4 + 2)(x^3 + 6x^2 + 5x + 4) + (x^3 - 10x^2 - 9x - 7)$. Factor $q(x)$ and $r(x)$ in the same way.

$q(x) = x^3 + 6x^2 + 5x + 4$	$r(x) = x^3 - 10x^2 - 9x - 7$
$k = 2$; $j = 2^{k-1} = 2$	$k = 2$ $\quad j = 2^{k-1} = 2$
$b = a_{2^{k-1}-1} - 1 = a_1 - 1 = 4$	$b = a_{2^{k-1}-1} - 1 = a_1 - 1 = -10$

$$x + 6$$
$$x^2 + 4\overline{\smash{\big)}\, x^3 + 6x^2 \mid + 5x + 4}$$
$$x^3 + 6x^2 \mid + 4x + 24$$
$$x - 20$$

$$x - 10$$
$$x^2 - 10\overline{\smash{\big)}\, x^3 - 10x^2 \mid - 9x - 7}$$
$$x^3 - 10x^2 \mid - 10x + 100$$
$$x - 107$$

Thus

$$q(x) = (x^2 + 4)(x + 6) + (x - 20)$$

and

$$r(x) = (x^2 - 10)(x - 10) + (x - 107).$$

So

$$p(x) = (x^4 + 2) \cdot [(x^2 + 4) \cdot (x + 6) + (x - 20)] + [(x^2 - 10) \cdot (x - 10) + (x - 107)].$$

Using this formula, evaluating $p(x)$ requires five multiplications: three that appear explicitly in the factorization and two to compute x^2 and x^4. Horner's method would have required seven. Observe, however, that ten additions (and subtractions) are done instead of seven.

ANALYSIS OF POLYNOMIAL EVALUATION WITH PREPROCESSING OF COEFFICIENTS

The number of operations done to evaluate $p(x)$ may be easily counted by considering the three steps used to describe the procedure:

1. Evaluate $q(x)$ and $r(x)$ recursively. (This suggests the use of a recurrence relation.)
2. Compute x^j. (The largest j used is 2^{k-1}. $x^2, x^4, x^8, \ldots,$ and $x^{2^{k-1}}$ may be computed by doing $k - 1$ multiplications.)
3. Multiply $(x^j + b)$ by $q(x)$ and add $r(x)$ (one multiplication and two additions).

Let $M(k)$ be the number of multiplications done to evaluate a monic polynomial of degree $2^k - 1$, *not* counting computing the powers of x (since they may be computed once and used as needed). Let $A(k)$ be the number of additions (and subtractions). Then:

$$M(1) = 0$$

$$M(k) = 2M(k - 1) + 1 \qquad \text{for } k > 1$$

and

$$A(1) = 1$$

$$A(k) = 2A(k - 1) + 2 \qquad \text{for } k > 1$$

By expanding $M(k)$ a few times we see that $M(k) = 4M(k - 2) + 2 + 1 = 8M(k - 3) + 4 + 2 + 1 = \sum_{i=0}^{k-2} 2^i = 2^{k-1} - 1$. The total number of multiplications, then, is $2^{k-1} - 1 + (k - 1)$, the last term for computing powers of x. Since $n = 2^k - 1$, the number of multiplications is $[(n + 1)/2] + \log (n + 1) - 2$, or roughly $(n/2) + \log n$. It is easy to show that $A(k) = (3n - 1)/2$. (The reader should check that these formulas describe the number of operations done in the example.)

Whether or not eliminating $(n/2) - \log n$ multiplications by doing $n/2$ extra additions is a timesaver, we have illustrated an important point: Lower bounds that have been obtained for a problem without preprocessing, in this case n */'s for evaluating a polynomial of degree n, may no longer be valid. The particular operations permitted in the preprocessing (for example, division of polynomials, as in this case, or finding roots of polynomials) can also affect the number of operations required. A lower bound of $\lceil n/2 \rceil$ */'s has been established for polynomial evaluation, allowing a variety of preprocessing operations.

The reader should do Exercise 5.5 now.

5.3   *Vector and Matrix Multiplication*

We begin by reviewing the well-known methods for multiplying matrices and vectors, noting the number of operations [multiplications and additions/subtractions ($\pm$'s)] used by these methods, and giving the known lower bounds in terms of multiplications and divisions (*/'s). Throughout this section we use capital letters for the names of vectors and matrices and the corresponding small letters for their components. The components are real numbers.

Let $V = (v_1, v_2, \ldots, v_n)$ and $W = (w_1, w_2, \ldots, w_n)$ be two n-vectors, that is, vectors with n components in each. The dot product of V and W, denoted $V \cdot W$, is defined as $V \cdot W = \Sigma_{i=1}^{n} v_i w_i$. Computing $V \cdot W$ as indicated by the definition requires n multiplications and $n - 1$ additions. It has been shown that even if one of the vectors is known in advance and some preprocessing of its components is permitted, at least n */'s are required in the worst case. Thus the straightforward computation of dot products is optimal.

Let A be an $m \times n$ matrix and let V be an n-vector. Let W be the product $A \cdot V$. By definition, the ith component of W is the dot product of the ith row of A with V. That is, for $1 \leq i \leq m$, $w_i = \Sigma_{j=1}^{n} a_{ij} v_j$. Computing $A \cdot V$ as indicated by the definition requires mn multiplications. This is known to be optimal. The number of additions done is $m(n - 1)$.

Let A be an $m \times n$ matrix, let B be an $n \times q$ matrix, and let C be the product of A and B. By definition, c_{ij} is the dot product of the ith row of A and the jth column of B. That is, $c_{ij} = \Sigma_{k=1}^{n} a_{ik} b_{kj}$ for $1 \leq i \leq m$ and $1 \leq j \leq q$. If the entries of C are computed by the usual matrix multiplication algorithm, i.e., as indicated by this formula, $m(n - 1) q$ additions and mnq multiplications are done. Much to the surprise of people studying the problem, attempts to prove that mnq */'s are required for matrix multiplication were unsuccessful, and eventually algorithms that do fewer */'s were sought and found. Two of these are presented here.

WINOGRAD'S MATRIX MULTIPLICATION

Suppose that the dot product of $V = (v_1, v_2, v_3, v_4)$ and $W = (w_1, w_2, w_3, w_4)$ is computed by the following formula:

$$V \cdot W = (v_1 + w_2) \cdot (v_2 + w_1) + (v_3 + w_4) \cdot (v_4 + w_3)$$

$$- v_1 v_2 - v_3 v_4 - w_1 w_2 - w_3 w_4 .$$

Observe that the last four multiplications involve only components of V or only components of W. There are only two multiplications that involve components of both vectors. (Also observe that the formula relies on the commutativity of multiplication; e.g. it uses the fact that $w_2 v_2 = v_2 w_2$. Hence it would not hold if multiplication of the components were not commutative; in particular, it would not hold if the components were matrices.)

Generalizing from the example, when n is even (say, $n = 2p$),

$$V \cdot W = \sum_{i=1}^{p} (v_{2i-1} + w_{2i}) \cdot (v_{2i} + w_{2i-1}) - \sum_{i=1}^{p} v_{2i-1} v_{2i} - \sum_{i=1}^{p} w_{2i-1} w_{2i}. \quad (5.1)$$

If n is odd, we let $p = \lfloor n/2 \rfloor$ and add the final term $v_n w_n$ to Eq. (5.1). In each summation, p, or $\lfloor n/2 \rfloor$, multiplications are done, so in all $3 \lfloor n/2 \rfloor$ multiplications are done. This is worse than the straightforward way of computing the dot product. Even if one of the vectors is known in advance and the second or third summation can be considered preprocessing, n multiplications would still be done. If *both* vectors are known in advance, then the whole computation could be thought of as preprocessing, thus eliminating the whole problem! So what has been gained by looking at a more complicated formula for the dot product?

Suppose we are to multiply the $m \times n$ matrix A by the $n \times q$ matrix B. Each row of A is involved in q dot products, one with each column of B, and each column of B is involved in m dot products, one with each row of A. Thus terms like the last two summations in Eq. (5.1) may be computed once for each row of A and column of B and used many times.

Algorithm 5.3 MATRIX MULTIPLICATION – WINOGRAD

Input: A, B, m, n, and q, where A and B are $m \times n$ and $n \times q$ matrices, respectively.

Output: $C = AB$.

1. $p \leftarrow \lfloor n/2 \rfloor$
2. ["Preprocess" rows of A.]

    for $i \leftarrow 1$ **to** m; ROWTERM$_i$ $\leftarrow \sum_{j=1}^{p} a_{i,2j-1} * a_{i,2j}$
3. ["Preprocess" columns of B.]

    for $i \leftarrow 1$ **to** q; COLMTERM$_i$ $\leftarrow \sum_{j=1}^{p} b_{2j-1,i} * b_{2j,i}$
4. [Compute entries of C.]

    for $i \leftarrow 1$ **to** m;

    for $j \leftarrow 1$ **to** q;

    $c_{ij} \leftarrow \sum_{k=1}^{p} (a_{i,2k-1} + b_{2k,j}) * (a_{i,2k} + b_{2k-1,j}) -$ ROWTERM$_i$ $-$ COLMTERM$_j$
5. [If n is odd, a final term is added to each entry of C.]

    if n is odd **then do**

            for $i \leftarrow 1$ **to** m;

            for $j \leftarrow 1$ **to** q;

              $c_{ij} \leftarrow c_{ij} + a_{in} * b_{nj}$

    end

ANALYSIS OF WINOGRAD'S ALGORITHM

Assume that n is even. (Analyzing Algorithm 5.3 for odd n is left as an exercise.) We count multiplications first. Step 2 does mp, step 3 does qp, and step 4 does mqp. The total, since $p = n/2$, is $(mnq/2) + (n/2)(q + m)$. If A and B are square matrices, both $n \times n$, then Winograd's algorithm does $(n^3/2) + n^2$ multiplications instead of the usual n^3. The difference is significant even for small n. Unfortunately, Winograd's algorithm does extra $\pm$'s. We count the $\pm$'s as follows:

Step 2: $m(p - 1)$

Step 3: $q(p - 1)$

Step 4: For each of the mq entries of C,

  $2p$ (the two pluses in each term of the summation)
  $+ p - 1$ (to add the terms in the summation)
  $+ 2$ (to subtract ROWTERM$_i$ and COLMTERM$_j$).

Thus step 4 does $mq(3p + 1)$ $\pm$'s, and the total, again assuming n is even, is $\frac{3}{2}mnq + (n/2)(m + q) + mq - m - q$. For square $(n \times n)$ matrices, where the comparison between algorithms is a little easier to see, Winograd's algorithm does $\frac{3}{2}n^3 + 2n^2 - 2n$ $\pm$'s instead of the usual $n^3 - n^2$.

Observe that there are fewer instructions that require incrementing and testing loop counters in Winograd's algorithm than there are in the usual method. On the other hand, Winograd's algorithm uses more complex subscripting and it requires fetching matrix entries more often than the usual method does. Exercise 5.8 and Program 1 explore these differences.

LOWER BOUNDS FOR MATRIX MULTIPLICATION

Winograd's algorithm shows that $m \times n$ and $n \times q$ matrices can be multiplied using fewer than mnq multiplications. How many $*/$'s are necessary? Is it $\Theta(mnq)$ or can the cubic term be eliminated? The best known lower bound is surprisingly low: mn, or n^2 for square matrices. We stated earlier that mn $*/$'s are necessary to multiply an $m \times n$ matrix by an n-vector. We would expect matrix multiplication to be at least as hard, hence to require at least as many $*/$'s, and Fig. 5.1 illustrates that this is true by showing that an algorithm to multiply matrices can be used to obtain a matrix vector product. (The two problems are the same, of course, if $q = 1$.) There is no matrix multiplication algorithm known that does only mn $*/$'s and it is not likely that one exists. However, there is an algorithm that, for large matrices, does significantly fewer multiplications *and* $\pm$'s than Winograd's.

STRASSEN'S MATRIX MULTIPLICATION

For the remainder of this section we assume the matrices to be multiplied are $n \times n$ square matrices, A and B. The key to Strassen's algorithm is a method for multiplying

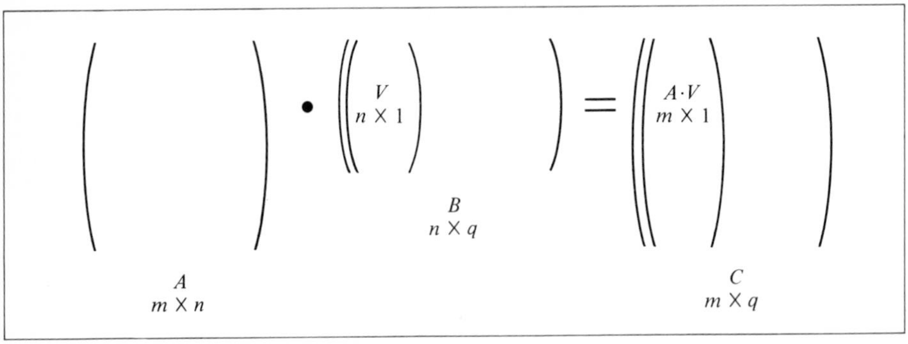

Figure 5.1
Lower bound for matrix multiplication.

2×2 matrices using seven multiplications instead of the usual eight. (Winograd's algorithm also uses eight.) For $n = 2$, first compute the following seven quantities, each of which requires exactly one multiplication:

$$x_1 = (a_{11} + a_{22}) \cdot (b_{11} + b_{22}), \qquad x_5 = (a_{11} + a_{12}) \cdot b_{22},$$

$$x_2 = (a_{21} + a_{22}) \cdot b_{11}, \qquad x_6 = (a_{21} - a_{11}) \cdot (b_{11} + b_{12}),$$

$$x_3 = a_{11} \cdot (b_{12} - b_{22}), \qquad x_7 = (a_{12} - a_{22}) \cdot (b_{21} + b_{22}). \quad (5.2)$$

$$x_4 = a_{22} \cdot (b_{21} - b_{11}),$$

Let $C = AB$. The entries of C are:

$$c_{11} = a_{11}b_{11} + a_{12}b_{21}, \qquad c_{12} = a_{11}b_{12} + a_{12}b_{22},$$

$$c_{21} = a_{21}b_{11} + a_{22}b_{21}, \qquad c_{22} = a_{21}b_{12} + a_{22}b_{22}.$$

They are computed as follows:

$$c_{11} = x_1 + x_4 - x_5 + x_7, \qquad c_{12} = x_3 + x_5,$$

$$c_{21} = x_2 + x_4, \qquad c_{22} = x_1 + x_3 - x_2 + x_6. \quad (5.3)$$

Thus 2×2 matrices can be multiplied using seven multiplications and 18 additions. It is critical to Strassen's algorithm that commutativity of multiplication is not used in the formulas in Eq. (5.2), so they can be applied to matrices whose components are also matrices. Let n be a power of two. Strassen's method consists of partitioning A and B each into four $n/2 \times n/2$ matrices as shown in Fig. 5.2 and multiplying them using the formulas in Eqs. (5.2) and (5.3); the formulas are used recursively to multiply the component matrices. Before considering extensions for the case when n is not a power of 2, we compute the number of multiplications and $\pm$'s done.

Suppose that $n = 2^k$ for some $k \geqslant 0$. Let $M(k)$ be the number of multiplications (of the underlying matrix components, i.e., real numbers) done by Strassen's method for $n \times n$ matrices. Then, since Eqs. (5.2) do seven multiplications of $2^{k-1} \times 2^{k-1}$ matrices,

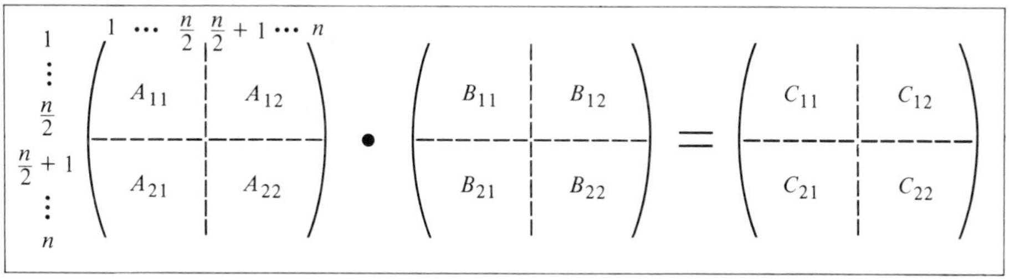

Figure 5.2
Partitioning for Strassen's matrix multiplication.

$$M(0) = 1$$
$$M(k) = 7M(k-1) \qquad \text{for } k > 0.$$

This recurrence relation is very easy to solve. $M(k) = 7^k$, and $7^k = 7^{\log n} = n^{\log 7} \approx n^{2.81}$. Thus the order of the number of multiplications is less than $\Theta(n^3)$.

Let $P(k)$ be the number of $\pm$'s done. Clearly $P(0) = 0$. There are 18 $\pm$'s in the formulas of Eqs. (5.2) and (5.3) so $P(1) = 18$. For $k \geqslant 1$, multiplying $2^k \times 2^k$ matrices requires 18 additions of $2^{k-1} \times 2^{k-1}$ matrices, plus all the $\pm$'s done by the seven matrix multiplications in Eqs. (5.2). So

$$P(0) = 0$$
$$P(k) = 18(2^{k-1})^2 + 7P(k-1) \qquad \text{for } k > 0.$$

Expanding the recurrence relation to see what the terms look like gives

$$\begin{aligned} P(k) &= 18(2^{k-1})^2 + 7P(k-1) = 18(2^{k-1})^2 + 7 \cdot 18(2^{k-2})^2 + 7^2 P(k-2) \\ &= 18(2^{k-1})^2 + 7 \cdot 18(2^{k-2})^2 + 7^2 \cdot 18(2^{k-3})^2 + 7^3 P(k-3). \end{aligned}$$

So

$$\begin{aligned} P(k) &= \sum_{i=0}^{k-1} 7^i \cdot 18(2^{k-i-1})^2 = 18 \cdot (2^k)^2 \sum_{i=0}^{k-1} \frac{7^i}{(2^{i+1})^2} \\ &= \frac{9}{2}(2^k)^2 \sum_{i=0}^{k-1} \left(\frac{7}{4}\right)^i = \frac{9}{2}(2^k)^2 \left[\frac{\left(\frac{7}{4}\right)^k - 1}{\frac{7}{4} - 1} \right] = 6 \cdot 7^k - 6 \cdot 4^k \\ &\approx 6n^{2.81} - 6n^2, \end{aligned}$$

which is also less than $\Theta(n^3)$.

If n is not a power of two some extension of Strassen's algorithm must be used and more work will be done. There are two simple approaches, but both can be very slow. These are to add extra rows and columns of zeroes to make the dimension a power of 2, or to use Strassen's formulas so long as the dimension of the matrices is even and then use the usual algorithm when the dimension is odd. Another, more complicated possibility is to modify the algorithm so that at each level of the recursion, if the matrices to be multiplied have odd dimension, one extra row and column are added. Strassen described a fourth strategy, one that combines the advantages of the first two. The matrices are embedded in larger ones with dimension $2^k m$, where $k = \lfloor \log n - 4 \rfloor$ and $m = \lfloor n/2^k \rfloor + 1$. Strassen's formulas are used recursively until the matrices to be multiplied are $m \times m$; then the usual method is applied. The total number of arithmetic operations done on the matrix entries will be less than $4.7 n^{\log 7}$.

Table 5.1 compares the numbers of arithmetic operations done by the three matrix multiplication methods for $n \times n$ matrices. For large n, Strassen's algorithm does fewer multiplications *and* fewer ±'s than either of the other methods. However, it is not a very good algorithm in practice. Because of the recursive nature of the algorithm, an implementation would require a lot of bookkeeping which will be very slow and/or complicated. The other, much simpler algorithms will be more efficient for moderate-sized n.

The major importance of Strassen's algorithm is theoretical: It broke the $\Theta(n^3)$ barrier for matrix multiplication *and* it broke the $\Theta(n^3)$ barrier for a number of other problems involving $n \times n$ matrices. These problems, which include matrix inversion, computing determinants, and solving systems of simultaneous linear equations, have well-known $\Theta(n^3)$ solutions, but since they can be reduced to matrix multiplication they too can be solved in $\Theta(n^{\log 7})$ time. It still remains for efficient algorithms with complexity less than $\Theta(n^3)$ to be found, and it still remains for lower bounds greater than n^2 to be established for matrix multiplication.

Table 5.1

Comparison of matrix multiplication methods for $n \times n$ matrices

	The usual algorithm	Winograd's algorithm	Strassen's algorithm
Multiplications	n^3	$\dfrac{n^3}{2} + n^2$	$7^k \approx n^{2.81}$, where $n = 2^k$
Additions/subtractions	$n^3 - n^2$	$\frac{3}{2}n^3 + 2n^2 - 2n$	$6 \cdot 7^k - 6 \cdot 4^k \approx 6n^{2.81} - 6n^2$, where $n = 2^k$
Total	$2n^3 - n^2$	$2n^3 + 3n^2 - 2n$	$4.7n^{\log 7} \approx 4.7n^{2.81}$ (n need not be a power of 2)

5.4   *The Fast Fourier Transform and Convolution*

Let U and V be n-vectors with components indexed from 0 to $n - 1$. The *convolution* of U and V, denoted $U \otimes V$, is, by definition, an n-vector W with components $w_i = \sum_{j=0}^{n-1} u_j v_{i-j}$, where $0 \leqslant i \leqslant n - 1$ and the indexes on the right-hand side are taken modulo n. For example, for $n = 5$,

$$w_0 = u_0 v_0 + u_1 v_4 + u_2 v_3 + u_3 v_2 + u_4 v_1$$

$$w_1 = u_0 v_1 + u_1 v_0 + u_2 v_4 + u_3 v_3 + u_4 v_2$$

$$\vdots$$

$$w_4 = u_0 v_4 + u_1 v_3 + u_2 v_2 + u_3 v_1 + u_4 v_0$$

The problem of computing the convolution of two vectors arises naturally and frequently in probability problems, engineering, and other areas. Symbolic polynomial multiplication, which will be examined in this section, is a convolution computation. An operator called the discrete Fourier transform (which will be defined later) can be used to compute convolutions and has a very large number of other applications. It is used in interpolation problems, in finding solutions of partial differential equations, in circuit design, and, very extensively, in signal processing.

The discrete Fourier transform of an n-vector and the convolution of two n-vectors may each be computed in a straightforward way using n^2 multiplications and fewer than n^2 additions. We will develop an algorithm to compute the discrete Fourier transform using fewer than $n \log n$ multiplications. This algorithm (which appears in the literature in many variations) is known as the fast Fourier transform, or FFT. We will then use the FFT to compute convolutions in $\Theta(n \log n)$ time. This saving of time is very valuable in the applications.

Throughout this section all matrix, array, and vector indexes will begin at 0. The complex roots of unity and some of their elementary properties are used in the FFT. The basic definitions and required properties are reviewed in the appendix at the end of Section 5.4. If the reader is not familiar with nth roots of unity, he or she should read the appendix before proceeding.

THE DISCRETE FOURIER TRANSFORM

The discrete Fourier transform transforms an n-vector with real components into a complex n-vector. For $n \geq 1$, let ω be a primitive nth root of 1, and let F_n be the $n \times n$ matrix with entries $f_{ij} = \omega^{ij}$, where $0 \leq i, j \leq n - 1$. The *discrete Fourier transform* of the n-vector $P = (p_0, p_1, \ldots, p_{n-1})$ is the product $F_n P$. The components of $F_n P$ are:

$$\omega^0 p_0 + \omega^0 p_1 + \cdots + \omega^0 p_{n-2} + \omega^0 p_{n-1}$$

$$\omega^0 p_0 + \omega p_1 + \cdots + \omega^{n-2} p_{n-2} + \omega^{n-1} p_{n-1}$$

$$\vdots$$

$$\omega^0 p_0 + \omega^i p_1 + \cdots + \omega^{i(n-2)} p_{n-2} + \omega^{i(n-1)} p_{n-1}$$

$$\vdots$$

$$\omega^0 p_0 + \omega^{n-1} p_1 + \cdots + \omega^{(n-1)(n-2)} p_{n-2} + \omega^{(n-1)(n-1)} p_{n-1}.$$

Rewritten in a slightly different form, the ith component is

$$p_{n-1}(\omega^i)^{n-1} + p_{n-2}(\omega^i)^{n-2} + \cdots + p_1 \omega^i + p_0.$$

Thus if we interpret the components of P as coefficients of the polynomial $p(x) = p_{n-1}x^{n-1} + p_{n-2}x^{n-2} + \cdots + p_1 x + p_0$, then the ith component is $p(w^i)$ and computing the discrete Fourier transform of P means evaluating the polynomial $p(x)$ at $\omega^0, \omega, \omega^2, \ldots, \omega^{n-1}$, i.e., at each of the nth roots of 1. We will approach the problem from this point of view. We will develop a recursive algorithm first and then examine it closely to remove the recursion. We assume that $n = 2^k$ for some $k \geqslant 0$. (Adjustments to the algorithm can be made if it is to be used for n not a power of 2.)

Recall that $\omega^{n/2} = -1$ and so for $0 \leqslant j \leqslant (n/2) - 1$, $\omega^{(n/2)+j} = -\omega^j$. Group the terms of $p(x)$ with even powers and the terms with odd powers as follows:

$$p(x) = \sum_{i=0}^{n-1} p_i x^i = \sum_{i=0}^{\frac{n}{2}-1} p_{2i} x^{2i} + x \cdot \sum_{i=0}^{\frac{n}{2}-1} p_{2i+1} x^{2i}.$$

Define

$$p_{\text{even}}(x) = \sum_{i=0}^{\frac{n}{2}-1} p_{2i} x^i \quad \text{and} \quad p_{\text{odd}}(x) = \sum_{i=0}^{\frac{n}{2}-1} p_{2i+1} x^i.$$

Then

$$p(x) = p_{\text{even}}(x^2) + x p_{\text{odd}}(x^2) \quad \text{and} \quad p(-x) = p_{\text{even}}(x^2) - x p_{\text{odd}}(x^2). \quad (5.4)$$

Equation (5.4) shows that to evaluate p at $1, \omega, \ldots, \omega^{(n/2)-1}, -1, -\omega, \ldots, -\omega^{(n/2)-1}$, it suffices to evaluate p_{even} and p_{odd} at $1, \omega^2, \ldots, (\omega^{(n/2)-1})^2$ and then do $n/2$ multiplications (for $x * p_{\text{odd}}(x^2)$) and n additions and subtractions. The polynomials p_{even} and p_{odd} have degree $(n/2) - 1$, or $2^{k-1} - 1$, and $1, \omega^2, \ldots, (\omega^{(n/2)-1})^2$ are the $(n/2)$th roots of 1, so p_{even} and p_{odd} can be evaluated recursively by the same scheme. Clearly, when the polynomial to be evaluated is a constant, there is no work to be done.

The recursive algorithm is as follows.

Algorithm 5.4 EVAL $(k, [p_0, p_1, \ldots, p_{2^k-1}], m, [\text{VAL}_0, \ldots, \text{VAL}_{2^k-1}])$

Input: $k, [p_0, p_1, \ldots, p_{2^k-1}], m$.

Output: The discrete Fourier transform of the vector $P = (p_0, p_1, \ldots, p_{2^k-1})$ stored in the array VAL.

Comment: The 2^k-th roots of 1: $\omega^0, \omega, \ldots, \omega^{2^k-1}$, are assumed to be in the array OMEGA in the order shown. EVAL would be called initially with $m = 1$. We use m to select roots from this array. In general, the set consisting of every mth entry, that is, $\omega^0, \omega^m, \omega^{2m}, \ldots$, is the set of $2^k/m$th roots of 1.

1.   **if** $k = 0$ **then** $\text{VAL}_0 \leftarrow \text{VAL}_1 \leftarrow p_0$
         else do
                  [Evaluate p_{even} at the 2^{k-1}th roots of 1.]
2.   EVAL $(k-1, [p_0, p_2, \ldots, p_{2^k-2}], 2m, [\text{VAL}'_0, \ldots, \text{VAL}'_{2^{k-1}-1}])$
                  [Evaluate p_{odd} at the 2^{k-1}-th roots of 1.]

3. $\qquad$ EVAL $\;(k-1,\;\;[p_1, p_3, \ldots, p_{2^k-1}],\;\;2m,\;\;[\mathrm{VAL}_0'', \ldots,$
    $\qquad\qquad$ $\mathrm{VAL}_{2^{k-1}-1}''])$
4. $\qquad\qquad$ **for** $j \leftarrow 0$ **to** $2^{k-1}-1$ **do** [evaluate $p(\omega^j)$ and $p(\omega^{2^{k-1}+j})$]
5. $\qquad\qquad\qquad$ XPODD $\leftarrow$ OMEGA$(mj) \cdot \mathrm{VAL}_j''$
6. $\qquad\qquad\qquad$ $\mathrm{VAL}_j \leftarrow \mathrm{VAL}_j' + \mathrm{XPODD}$ [value of $p(\omega^j)$]
7. $\qquad\qquad\qquad$ $\mathrm{VAL}_{2^{k-1}+j} \leftarrow \mathrm{VAL}_j' - \mathrm{XPODD}$ [value of $p(\omega^{2^{k-1}+j})$]
    $\qquad\qquad$ **end**
    $\qquad$ **end**

The recursiveness of the algorithm makes it easy to find a recurrence relation for the number of operations done. We will count the arithmetic operations done on components of P and roots of 1. Let $M(k)$, $A(k)$, and $S(k)$ be the number of multiplications, additions, and subtractions, respectively, done by EVAL to compute the discrete Fourier transform of a 2^k-vector. The three operations are done, one each, in lines 5, 6, and 7, so $M(k) = A(k) = S(k)$. We solve for $M(k)$. Clearly, by line 1,

$$M(0) = 0.$$

From lines 2, 3, and 5 we see that

$$M(k) = 2^{k-1} + 2M(k-1),$$

where the first term, 2^{k-1}, counts the multiplications in line 5, and the second term, $2M(k-1)$, counts multiplications done by the recursive calls to EVAL to compute the values in the arrays VAL$'$ and VAL$''$. It is easy to see that $M(k) = 2^{k-1}k$. Thus $M(k) = A(k) = S(k) = 2^{k-1}k$, or $(n/2) \log n$. Since the operations are done on complex numbers, this result should be multiplied by a small constant to reflect the fact that each complex operation requires several ordinary ones. (See Exercise 5.12.)

As presented above, the algorithm would require a lot of extra time and space for the bookkeeping necessitated by the recursion. Yet the breakdown of the polynomial seems systematic enough that we should be able to obtain a scheme for carrying out the same computation "from the bottom up" without using a recursive program. To help suggest the pattern of the computation, an example is presented in a tree diagram in Fig. 5.3. The levels of the tree correspond to the levels of recursion, but the recursion could be eliminated if we start the computation at the leaves. The leaves are the components of the vector P permuted in a particular way. Determining the correct permutation, π_k, is the key to constructing an efficient implementation of the evaluation scheme. π_k will be given (and its correctness proved) after the presentation of the nonrecursive algorithm. The reader is invited to try to determine how to define π_k before proceeding.

Observe that at each level of the tree the same number of values are to be computed: 2^k, since at level ℓ there are 2^ℓ nodes, or polynomials, to be evaluated at $2^{k-\ell}$ roots of unity. Since the values computed at one level are needed only for the computation of two values in the next level, one array, VAL, with 2^k entries suffices to store the results of the computations. The variables used in the algorithm play the following roles: ℓ is the level number, NVAL is the number of values to be computed

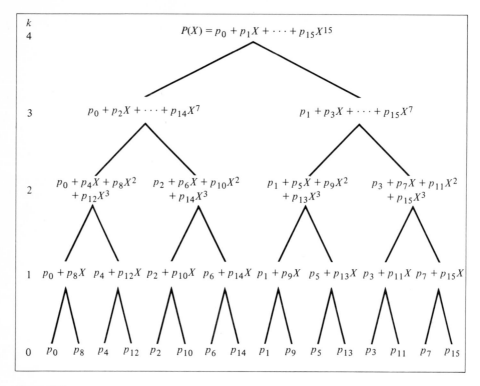

Figure 5.3
Polynomial evaluation at roots of unity. For a polynomial p at any internal node, the left child is p_{even} and the right child is p_{odd}.

at each node at level ℓ, t is the index in VAL for the first of these values for a particular node, and j counts off the pairs of values to be computed for that node; m is used as in EVAL to pick out the correct entry from OMEGA. Figure 5.4 illustrates how two values at a node at level ℓ are computed using one value from each of its children. The diagram may help clarify the algorithm's indexing.

Algorithm 5.5   THE FAST FOURIER TRANSFORM

Input: The n-vector $P = (p_0, p_1, \ldots, p_{n-1})$, where $n = 2^k$ for some $k > 0$.

Output: VAL, the discrete Fourier transform of P.

Comment: OMEGA is an array containing the nth roots of 1: ω^0, $\omega, \ldots, \omega^{n-1}$. VAL is initialized to contain the values for level $k-1$, not the leaves, in the tree of Fig. 5.3.

[To initialize VAL, evaluate polynomials of degree $1 = 2^1 - 1$ at square roots of 1.]
1. **for** $t \leftarrow 0$ **to** $n - 2$ **by** 2 **do**
2. $VAL(t) \leftarrow p_{\pi_k(t)} + p_{\pi_k(t+1)}$

3. $\text{VAL}(t + 1) \leftarrow p_{\pi_k(t)} - p_{\pi_k(t+1)}$

4.   **end**

[NVAL is the number of points at which each polynomial at the current level is evaluated. The levels are indexed by ℓ.]

5. $m \leftarrow \dfrac{n}{2}$; $\text{NVAL} \leftarrow 2$

6.   **for** $\ell \leftarrow k - 2$ **to** 0 **by** -1 **do**

7. $m \leftarrow \dfrac{m}{2}$; $\text{NVAL} \leftarrow 2 \cdot \text{NVAL}$

8.   **for** $t \leftarrow 0$ **to** $(2^{\ell} - 1)\text{NVAL}$ **by** NVAL;

9.   **for** $j \leftarrow 0$ **to** $\dfrac{\text{NVAL}}{2} - 1$ **do**

10. $\text{XPODD} \leftarrow \text{OMEGA}\,(mj) \cdot \text{VAL}\left(t + \dfrac{\text{NVAL}}{2} + j\right)$

11. $\text{VAL}\left(t + \dfrac{\text{NVAL}}{2} + j\right) \leftarrow \text{VAL}(t + j) - \text{XPODD}$

12. $\text{VAL}(t + j) \leftarrow \text{VAL}(t + j) + \text{XPODD}$

13.   **end**

14.   **end**

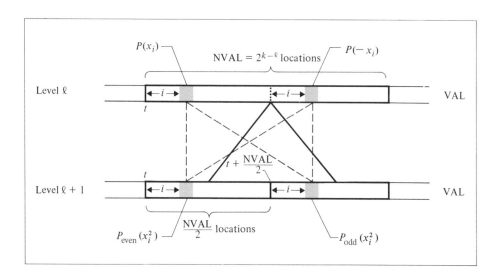

Figure 5.4

Illustration for the FTT. At the node shown in level ℓ, the polynomial p is to be evaluated at x_0, $x_1, \ldots, x_{2^{k-(\ell+1)-1}}$, $-x_0$, $-x_1, \ldots, -x_{2^{k-(\ell+1)-1}}$, where $P(x_i) = P_{\text{even}}(x_i^2) + x_i P_{\text{odd}}(x_i^2)$ and $P(-x_i) = P_{\text{even}}(x_i^2) - x_i P_{\text{odd}}(x_i^2)$. The diagram shows which values from the previous levels are used to get $P(x_i)$ and $P(-x_i)$.

An analysis of the number of operations done by the FFT gives a result only slightly different from that obtained for EVAL. Lines 10, 11, and 12, which do one multiplication, one addition, and one subtraction, respectively, are in a triply nested loop. It is easy to verify that $NVAL = 2^{k-\ell}$ so the ranges of the loop indexes indicate that the number of each operation done in these lines is

$$\sum_{\ell=0}^{k-2} 2^\ell \frac{NVAL}{2} = \sum_{\ell=0}^{k-2} 2^\ell 2^{k-\ell-1}$$

$$= \sum_{\ell=0}^{k-2} 2^{k-1}$$

$$= (k-1)2^{k-1}, \quad \text{or} \quad \frac{n}{2}\log\frac{n}{2}.$$

Lines 2 and 3 do $n/2$ additions and $n/2$ subtractions, so the total is $\frac{3}{2}n\log(n/2) + n$ complex arithmetic operations. We claim that the permutation π_k can be computed easily enough so that the running time of the FFT is $\Theta(n\log n)$.

Note that the FFT allows us to evaluate a polynomial of degree $n-1$ at n distinct points at a cost of only $(n/2)\log(n/2)$ complex multiplications. The lower bound on polynomial evaluation given in Section 5.2 suggests that this is not possible for n arbitrary points. The speed of the FFT derives from its use of some of the properties of roots of unity.

Now, what is π_k? Let t be an integer between 0 and $n-1$, where $n = 2^k$. Then t can be represented in binary by $[b_0 b_1 \ldots b_{k-1}]$, where each b_j is 0 or 1. Let $rev_k(t)$ be the number represented by these bits in reverse order, i.e., by $[b_{k-1} \ldots b_1 b_0]$. We claim that $\pi_k(t) = rev_k(t)$. Lemma 5.2 describes the values computed by the FFT using $\pi_k = rev_k$ and is used in Theorem 5.3 to establish the correctness of the algorithm, thus also establishing the correctness of this choice of π_k. The proof of the lemma follows the theorem.

Lemma 5.2 Let π_k in Algorithm 5.5 be rev_k. The following statements hold for $\ell = k-1$ after execution of line 5 and for each ℓ such that $k-2 \geqslant \ell \geqslant 0$ at the end of the loop in lines 6–14.

1. $m = 2^\ell$ and $NVAL = 2^{k-\ell}$.

2. For $t = r2^{k-\ell}$ where $0 \leqslant r \leqslant 2^\ell - 1$, the locations $VAL(t), \ldots, VAL(t + NVAL - 1)$ contain the values of $P^{t,\ell}$ evaluated at the $2^{k-\ell}$-th roots of 1, where $P^{t,\ell}$ is the polynomial of degree $2^{k-\ell} - 1$ with coefficients $c_j^{t,\ell} = p_{2\ell j + rev_k(t)}$ for $0 \leqslant j \leqslant 2^{k-\ell} - 1$.

Theorem 5.3 Algorithm 5.5 computes the values of

$$p(x) = p_0 + p_1 x + \cdots + p_{n-2} x^{n-2} + p_{n-1} x^{n-1}$$

at the nth roots of 1 — i.e., it computes the discrete Fourier transform of $P = (p_0, p_1, \ldots, p_{n-1})$.

Proof. Let $\ell = 0$ in Lemma 5.2. Then the only value for t is 0 and the lemma says that VAL(0), ..., VAL($2^k - 1$) contain the values of $P^{0,0}$ at the 2^k-th roots of 1. The coefficients of $P^{0,0}$ are $c_j^{0,0} = p_{2^0 j + \text{rev}(0)} = p_j$ for $0 \leqslant j \leqslant 2^k - 1$, so $P^{0,0}$ is the polynomial p. $\qquad\square$

Proof of Lemma 5.2. We prove the lemma by induction on ℓ where ℓ ranges from $k - 1$ down to 0. Let $\ell = k - 1$ for the basis. (1) is clearly true. (2) states that t ranges from 0 to $2^k - 2$ (i.e., $n - 2$) in steps of 2, and that for each t, VAL(t) and VAL($t + 1$) contain $P^{t, k-1}(1)$ and $P^{t, k-1}(-1)$, where the coefficients of $P^{t, k-1}$ are $c_0^{t, k-1} = p_{2^{k-1} 0 + \text{rev}(t)} = p_{\text{rev}(t)}$ and $c_1^{t, k-1} = p_{2^{k-1} + \text{rev}(t)} = p_{\text{rev}(t+1)}$ (see Lemma 5.4 below). That is, $p^{t, k-1}(x) = p_{\text{rev}(t)} + p_{\text{rev}(t+1)} x$. This corresponds exactly to the initial values assigned to VAL in lines 1–4.

Now suppose that $0 \leqslant \ell < k - 1$ and (1) and (2) hold for $\ell + 1$. It follows easily that (1) holds for ℓ. Note that ω^{2^ℓ} and $\omega^{2^{\ell+1}}$ are primitive $2^{k-\ell}$-th and $2^{k-(\ell+1)}$-th roots of 1, respectively. Using the induction assumption, we see that for $0 \leqslant i \leqslant$ NVAL$/2 - 1$ lines 10–12 compute

$$\text{XPODD} \leftarrow (\omega^{2^\ell})^i P^{t + (\text{NVAL}/2), \ell+1}((\omega^{2^{\ell+1}})^i),$$

$$\text{VAL}\left(t + \frac{\text{NVAL}}{2} + i\right) \leftarrow P^{t, \ell+1}((\omega^{2^{\ell+1}})^i) - (\omega^{2^\ell})^i P^{t + (\text{NVAL}/2), \ell+1}((\omega^{2^{\ell+1}})^i),$$

$$\text{VAL}(t + i) \leftarrow P^{t, \ell+1}((\omega^{2^{\ell+1}})^i) + (\omega^{2^\ell})^i P^{t + (\text{NVAL}/2), \ell+1}((\omega^{2^{\ell+1}})^i).$$

Thus $P^{t, \ell}(x) = P^{t, \ell+1}(x^2) + x P^{t+(\text{NVAL}/2), \ell+1}(x^2)$ and VAL(0), ..., VAL($t +$ NVAL $- 1$) contain $P^{t, \ell}$ evaluated at

$$(\omega^{2^\ell})^0, \omega^{2^\ell}, \ldots, (\omega^{2^\ell})^{2^{k-\ell-1}}, -(\omega^{2^\ell})^0, -\omega^{2^\ell}, \ldots, -(\omega^{2^\ell})^{2^{k-\ell-1}}$$

i.e., at the $2^{k-\ell}$-th roots of 1. The coefficients of $P^{t, \ell}$ are derived as follows for $0 \leqslant j \leqslant 2^{k-\ell} - 1$:

$$c_j^{t, \ell} = c_{j/2}^{t, \ell+1} \qquad \text{for even } j$$

$$c_j^{t, \ell} = c_{(j-1)/2}^{t+(\text{NVAL}/2), \ell+1} \qquad \text{for odd } j.$$

Therefore, using the induction hypothesis, for even j,

$$c_j^{t, \ell} = c_{j/2}^{t, \ell+1} = p_{2^{\ell+1}(j/2) + \text{rev}(t)} = p_{2^\ell j + \text{rev}(t)}$$

as required. For odd j,

$$c_j^{t, \ell} = c_{(j-1)/2}^{t+(\text{NVAL}/2), \ell+1}$$

$$= p_{2^{\ell+1}(j-1)/2 + \text{rev}(t + 2^{k-(\ell+1)})}$$

$$= \text{(by Lemma 5.4) } p_{2^\ell(j-1) + \text{rev}(t) + 2^\ell} = p_{2^\ell j + \text{rev}(t)},$$

also as required. $\qquad\square$

The proof used the following lemma, whose proof is left as an exercise.

Lemma 5.4 For $k, a, b > 0$, $b \leqslant k$, and $a + 2^{k-b} < 2^k$, if a is a multiple of 2^{k-b+1}, then $\text{rev}_k(a + 2^{k-b}) = \text{rev}_k(a) + 2^{b-1}$.

CONVOLUTION

To motivate the convolution computation we will examine the problem of symbolic polynomial multiplication. Suppose we are given the coefficient vectors $P = (p_0, p_1, \ldots, p_{m-1})$ and $Q = (q_0, q_1, \ldots, q_{m-1})$ for the polynomials $p(x) = p_{m-1}x^{m-1} + p_{m-2}x^{m-2} + \cdots + p_1 x + p_0$ and $q(x) = q_{m-1}x^{m-1} + q_{m-2}x^{m-2} + \cdots + q_1 x + q_0$. The problem is to find the vector $R = (r_0, r_1, \ldots, r_{2m-1})$ of coefficients of the product polynomial $r(x) = p(x) \cdot q(x)$. The coefficients of r are given by the following formula: $r_i = \sum_{j=0}^{i} p_j q_{i-j}$ for $0 \leqslant i \leqslant 2m - 1$ with p_k and q_k taken as 0 for $k > m - 1$. (Note that $n_{2m-1} = 0$ since r has degree $2m - 2$; it is included as a convenience.) R is very much like the convolution of P and Q. Let $\bar{P}$ and $\bar{Q}$ be the $2m$-vectors obtained by adding m zeroes to P and Q, respectively. Then $R = \bar{P} \otimes \bar{Q}$. So our investigation of polynomial multiplication should lead to a convolution algorithm.

Consider the following outline for polynomial multiplication.

1. Evaluate $p(x)$ and $q(x)$ at $2m$ points: $x_0, x_1, \ldots, x_{2m-1}$.

2. Multiply pointwise to find the values of $r(x)$ at these $2m$ points; i.e., compute $r(x_i) = p(x_i) \cdot q(x_i)$ for $0 \leqslant i \leqslant 2m - 1$.

3. Find the coefficients of the unique polynomial of degree $2m - 2$ which passes through the points $\{(x_i, r(x_i)) : 0 \leqslant i \leqslant 2m - 1\}$. (It is a well-known theorem that the coefficients of a polynomial of degree d can be determined if the values of the polynomial are known at $d + 1$ points.)

If the points $x_0, \ldots, x_{2m-1}$ were chosen arbitrarily, the method outlined would require much more work than a straightforward computation of $\bar{P} \otimes \bar{Q}$, but the FFT can evaluate p and q at the $(2m)$th roots of 1 very efficiently (assume that m is a power of 2). So step 1 can be done in $\Theta(m \log m)$ time. Step 2 requires only $2m$ multiplications. How do we carry out step 3?

Let ω be a primitive $(2m)$th root of 1, and for $0 \leqslant j \leqslant 2m - 1$ let $v_j = r(\omega^j) = p(\omega^j) \cdot q(\omega^j)$. The coefficients of r may be found by solving the following set of simultaneous linear equations for $r_0, r_1, \ldots, r_{2m-1}$.

$$r_0 + r_1 \omega^0 + \cdots + r_{2m-2}(\omega^0)^{2m-2} + r_{2m-1}(\omega^0)^{2m-1} = v_0$$
$$r_0 + r_1 \omega + \cdots + r_{2m-2}(\omega)^{2m-2} + r_{2m-1}(\omega)^{2m-1} = v_1$$
$$\vdots$$
$$r_0 + r_1 \omega^{2m-1} + \cdots + r_{2m-2}(\omega^{2m-1})^{2m-2} + r_{2m-1}(\omega^{2m-1})^{2m-1} = v_{2m-1}$$

$$(5.5)$$

If r had been evaluated at $2m$ arbitrary points, an $\Theta(m^3)$ algorithm such as Gaussian elimination might be used to solve the equations. Again, we take advantage of the fact that the points are roots of unity to obtain an $\Theta(m \log m)$ algorithm. The equations in (5.5) can be written as a matrix equation $F_{2m}R = V$, where V is the vector $(v_0, v_1, \ldots, v_{2m-1})$. Thus $\bar{P} \otimes \bar{Q} = R = F_{2m}^{-1} V = F_{2m}^{-1}(F_{2m}\bar{P} * F_{2m}\bar{Q})$, where $*$ denotes pointwise multiplication. Three problems remain: to show that F_n is in fact invertible for all $n > 0$, to show that the formula $U \otimes V = F_n^{-1}(F_n U * F_n V)$ holds for arbitrary n-vectors U and V, and to find an efficient way to compute the inverse transform. The formula for $U \otimes V$ does not follow immediately from the formula for R because $\bar{P}$ and $\bar{Q}$ have the property that half of their components are zero.

Lemma 5.5 For $n > 0$, F_n is invertible and the (i, j)th entry of its inverse is $(1/n)\omega^{-ij}$ for $0 \leqslant i, j \leqslant n - 1$.

Proof. Let $\tilde{F}_n$ be the matrix which the lemma claims is F_n^{-1}. We show that $F_n \tilde{F}_n = I$; $\tilde{F}_n F_n = I$ similarly.

$$(F_n \tilde{F}_n)_{ij} = \sum_{k=0}^{n-1} \omega^{ik} \frac{1}{n} \omega^{-kj} = \frac{1}{n} \sum_{k=0}^{n-1} (\omega^{i-j})^k.$$

For nondiagonal entries, i.e., for $i \neq j$, $(F_n \tilde{F}_n)_{ij} = 0$ by Property 5.1 for roots of unity since $0 < |i - j| < n$ (see the appendix at the end of this section). For diagonal entries, i.e., for $i = j$,

$$(F_n \tilde{F}_n)_{ij} = \frac{1}{n} \sum_{k=0}^{n-1} (\omega^0)^k = \frac{1}{n} \sum_{k=0}^{n-1} 1 = 1. \qquad \square$$

Theorem 5.6 Let U and V be n-vectors. Then $U \otimes V = F_n^{-1}(F_n U * F_n V)$, where $*$ denotes pointwise multiplication.

Proof. We show that $F_n(U \otimes V) = F_n U * F_n V$. For $0 \leqslant i \leqslant n - 1$, the ith component of $F_n U * F_n V$ is

$$\left(\sum_{j=0}^{n-1} \omega^{ij} u_j \right) \left(\sum_{k=0}^{n-1} \omega^{ik} v_k \right) = \sum_{j=0}^{n-1} \sum_{k=0}^{n-1} u_j v_k \, \omega^{i(j+k)}.$$

The tth component of $U \otimes V$ is $\sum_{j=0}^{n-1} u_j v_{t-j}$ where subscripts are taken modulo n. So the ith component of $F_n(U \otimes V)$ is

$$\sum_{t=0}^{n-1} \left(\omega^{it} \sum_{j=0}^{n-1} u_j v_{t-j} \right) = \sum_{j=0}^{n-1} \sum_{t=0}^{n-1} u_j v_{t-j} \omega^{it}.$$

Let $k = t - j \pmod{n}$ in the inner summation. For each j, since t ranges from 0 to $n - 1$, k will also range from 0 to $n - 1$, although in a different order. Also for any p,

$\omega^p = \omega^{p(\bmod\ n)}$, so the ith component of $F_n(U \otimes V) = \sum_{j=0}^{n-1} \sum_{k=0}^{n-1} u_j v_k \omega^{i(j+k)}$ which is exactly the ith component of $F_n U * F_n V$.   □

Lemma 5.5 indicates that the matrix F_n^{-1} is not very much different from F_n. The entries of nF_n^{-1} are ω^{-ij}. Its rows are the rows of F_n arranged in a different order. Specifically, since $\omega^{n-i} = \omega^{-i}$, for $1 \leqslant i \leqslant n-1$ the ith row of F_n is the $(n-i)$th row of nF_n^{-1}. Row 0 is the same for both matrices. Thus the inverse discrete Fourier transform of an n-vector A may be computed as follows.

Algorithm 5.6 INVERSE DISCRETE FOURIER TRANSFORM

Input: A, n, where A is an n-vector; n is a power of 2.

Output: The vector $B = (b_0, b_1, \ldots, b_{n-1})$, where $B = F_n^{-1} A$.

1.   Compute $F_n A$ using Algorithm 5.5 and leaving the results in VAL.
2. $b_0 \leftarrow \text{VAL}(0)/n$
      for $i \leftarrow 1$ to $n-1$; $b_i \leftarrow \text{VAL}(n-i)/n$

ANALYSIS

The FFT does $(n/2)\log(n/2)$ complex multiplications so Algorithm 5.6 does $(n/2)$ $\log(n/2) + n$ complex $*/$'s and both run in $\Theta(n \log n)$ time. Computing the convolution of two n-vectors using the FFT takes $n \log(n/2) + 2n = n\log n + n$ $*/$'s.

APPENDIX: COMPLEX NUMBERS AND ROOTS OF UNITY

C, the field of complex numbers, is obtained by adjoining i, the square root of -1, to the field of real numbers R. Thus $C = R(i) = \{a + bi: a, b \in R\}$. If $z = a + bi$, a is called the real part of z and b the imaginary part. Let $z_1 = a_1 + b_1 i$ and $z_2 = a_2 + b_2 i$. Then, by definition, $z_1 + z_2 = (a_1 + a_2) + (b_1 + b_2)i$ and $z_1 \cdot z_2 = (a_1 a_2 - b_1 b_2) + (a_1 b_2 + b_1 a_2)i$. A complex number may be represented as a vector in a plane using the real and imaginary parts for the Cartesian coordinates. The geometric interpretation of multiplication of complex numbers is more easily seen by using polar coordinates, r and θ, where r is the length of the vector and θ is the angle (measured in radians) that it subtends with the horizontal, or real, axis. (See Fig. 5.5.) The product of two complex numbers (r_1, θ_1) and (r_2, θ_2) is $(r_1 r_2, \theta_1 + \theta_2)$. An example is given in Fig. 5.6.

C is an algebraically closed field; that means that every polynomial of degree n with coefficients in C has n roots. Therefore, $x^n - 1$ has n roots, which are called the *nth roots of unity*. The polar coordinates of 1 are $(1, 0)$. To find a root (r, θ) of $x^n - 1$ we solve the equation $(r^n, n\theta) = (1, 0)$. Since r is real and nonnegative, r must be 1, so all roots of unity are represented by vectors of unit length. Since $n\theta = 0$, $\theta = 0$ so we have found that $(1, 0) -$ i.e., $1 -$ is a solution; hardly a surprise. To find the other roots of unity we use the fact that an angle of 0 radians is equivalent to an angle of $2\pi j$ radians for any integer j. The n distinct roots are $\{(1, 2\pi j/n) : 0 \leqslant j \leqslant$

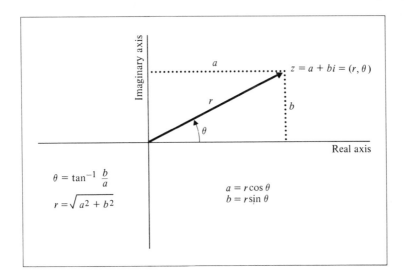

Figure 5.5
The relationship between Cartesian and polar coordinates for complex numbers.

$n - 1\}$. The vectors representing these numbers slice the unit circle into n equal pie slices as shown in Fig. 5.7.

If ω is an nth root of 1, then ω^k is also an nth root of 1 since $(\omega^k)^n = \omega^{nk} = (\omega^n)^k = 1^k = 1$. If ω is an nth root of 1 and $1, \omega, \omega^2, \ldots, \omega^{n-1}$ are all distinct, then ω is called a *primitive nth root of unity*. One primitive nth root of unity is $(1, 2\pi/n)$, or $\cos(2\pi/n) + i\sin(2\pi/n)$. The following properties are used in Section 5.4.

Property 5.1 For $n \geqslant 2$, the sum of all the nth roots of 1 is zero. Also, if ω is a primitive nth root of 1 and n does not divide c, then $\sum_{j=0}^{n-1} (\omega^c)^j = 0$.

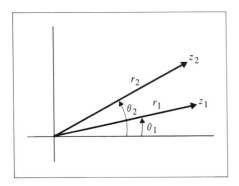
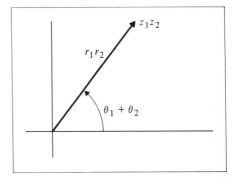

Figure 5.6
Multiplication of complex numbers.

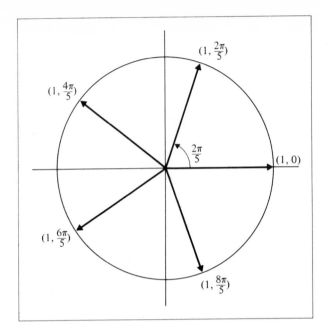

Figure 5.7
Fifth roots of unity (polar coordinates).

Proof. Let ω be a primitive nth root of 1. Then $\omega^0, \omega, \omega^2, \ldots, \omega^{n-1}$ are all of the nth roots of 1. Their sum is

$$\sum_{j=0}^{n-1} \omega^j = \frac{\omega^n - 1}{\omega - 1} = \frac{1 - 1}{\omega - 1} = 0.$$

The second statement is proved similarly.

Property 5.2 If n is even and ω is a primitive nth root of 1, then

1. ω^2 is a primitive $(n/2)$th root of 1, and

2. $\omega^{n/2} = -1$.

Proof. Exercise 5.10.

5.5   *Exercises*

Section 5.2: Evaluating Polynomial Functions

5.1.   Any polynomial $p(x) = a_n x^n + a_{n-1} x^{n-1} + \cdots + a_1 x + a_0$ may be factored: $a_n(x - r_1)(x - r_2), \ldots, (x - r_n)$, where $r_1, \ldots, r_n$ are the roots of p. Could this factorization be used as the basis of an algorithm to evaluate $p(x)$? How, or why not?

5.2. Write out the factorization that would be used to evaluate $p(x) = x^7 + 6x^6 - 7x^5 + 12x^4 + 2x^2 - 3x - 8$ by
    a) Horner's method;
    b) preprocessing coefficients.

5.3. What modifications or additions must be made to the procedure for evaluating polynomials with preprocessing of coefficients so that it will work for nonmonic polynomials? How many multiplications and/or divisions are done by the extended algorithm?

5.4. Suppose that $A(1) = 1$ and for $k > 1$, $A(k) = 2A(k - 1) + 2$. Show that $A(k) = (3n - 1)/2$.

5.5. Using the terminology of the first paragraph of "Preprocessing of Coefficients" in Section 5.2, what are I, I', A_I, and the algorithm that gives A_I from I for the problem of evaluating a polynomial with preprocessing of coefficients by the method described in Section 5.2?

Section 5.3: Vector and Matrix Multiplication

5.6. At the beginning of Section 5.3 we stated that computing the dot product of of two n-vectors with real components requires at least n */'s. Show by writing an algorithm that if one of the vectors, say V, always has integer components, fewer */'s suffice to compute the dot product of V and W. How many */'s are required?

5.7. Compute exactly the number of multiplications and additions done by Algorithm 5.3 when n is odd.

5.8. Let A and B be $n \times n$ matrices that are to be multiplied and suppose that a matrix entry must be fetched from storage each time it is used in the computation. How many times is each entry of A and B fetched to compute AB
    a) By the usual algorithm?
    b) By Winograd's algorithm (for n even)?

5.9. Prove that Strassen's algorithm, using the fourth modification described toward the end of Section 5.3, does less than $4.7n^{\log 7}$ arithmetic operations on the matrix entries, whether or not n is a power of 2.

Section 5.4: The Fast Fourier Transform and Convolution

5.10. Prove Property 5.2 for roots of unity. Why are the restrictions "$n \geq 2$" and "n does not divide c" needed in Property 5.1?

5.11. Let $p(x) = p_7 x^7 + p_6 x^6 + \cdots + p_1 x + p_0$. Carry out the steps of EVAL on p to show how it evaluates p at the 8th roots of 1: $1, \omega, i, i\omega, -1, -\omega, -i, -i\omega$.

5.12. Suppose you are given the real and imaginary parts of two complex numbers. Show that the real and imaginary parts of their product may be computed using only three multiplications.

5.13. Prove Lemma 5.4.

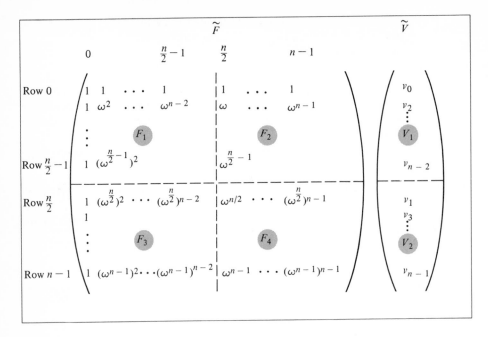

Figure 5.8

5.14. Let $n = 2^k$ for some $k > 0$, let ω be a primitive nth root of 1, and let V be an n-vector. This problem describes the FFT (recursively) from the point of view of the matrix-vector product $F_n V$ rather than as polynomial evaluation. The reader should note the correspondence of various steps of this algorithm with steps of EVAL.

Let $\widetilde{F}_n$ be the $n \times n$ matrix obtained from F_n by putting all the even indexed columns before the odd indexed columns, and let $\widetilde{V}$ have all the even indexed components of V precede all the odd indexed components; i.e., for $0 \leqslant j \leqslant (n/2) - 1$, $\tilde{f}_{ij} = \omega^{i \cdot 2j}$, $\tilde{f}_{i,j+(n/2)} = \omega^{i(2j+1)}$, $\tilde{v}_j = v_{2j}$, and $\tilde{v}_{j+(n/2)} = v_{2j+1}$. Partition $\widetilde{F}_n$ into four $(n/2) \times (n/2)$ matrices F_1, F_2, F_3, and F_4, and partition $\widetilde{V}$ into two $(n/2)$ - vectors V_1 and V_2 as shown in Fig. 5.8. Now

$$\widetilde{F}_n \widetilde{V} = \begin{pmatrix} F_1 V_1 + F_2 V_2 \\ F_3 V_1 + F_4 V_2 \end{pmatrix}.$$

Prove the following statements.
a) $F_n V = \widetilde{F}_n \widetilde{V}$.
b) $F_1 = F_3$.
c) $F_4 = -F_2$.
d) $F_2 = DF_1$, where D is an $(n/2) \times (n/2)$ diagonal matrix with $d_{ij} = \omega^i$ for $0 \leqslant i \leqslant (n/2) - 1$.

e) F_1 has entries $f_{ij} = x^{ij}$, where x is a primitive $(n/2)$th root of 1.

Thus
$$F_n V = \begin{pmatrix} F_1 V_1 + D F_1 V_2 \\ F_1 V_1 - D F_1 V_2 \end{pmatrix}$$

and the computation may be carried out by recursively computing the discrete Fourier transform of V_1 and V_2, both $(n/2)$-vectors.

f) Derive recurrence relations for the number of multiplications, additions, and subtractions that would be done by the algorithm described here. (Note that the product $D(F_1 V_2)$ may be done as a component-wise product of $\bar{D} = (1, \omega, \ldots, \omega^{(n/2)-1})$ with $F_1 V_2$, requiring $n/2$ multiplications.) Compare your recurrence relations with those obtained for EVAL.

PROGRAMS

1. Write and debug efficient assembly language subroutines for Winograd's matrix multiplication algorithm and for the usual algorithm. Using the instruction timing tables for your computer, compute the actual time required by each program to multiply two $n \times n$ matrices.

2. Implement the FFT (Algorithm 5.5). Make the computation of rev_k and the other bookkeeping as efficient as you can.

NOTES AND REFERENCES

The lower bounds given in Sections 5.2 and 5.3 for polynomial evaluation, with or without preprocessing of coefficients, and for vector-matrix products are established in Reingold and Stocks (1972) and in Winograd (1970). Winograd's matrix multiplication algorithm also appears in the latter. Winograd's proofs use field theory. Reingold and Stocks use simpler arguments such as that in the proof of Theorem 5.1.

Strassen's matrix multiplication algorithm was presented in Strassen (1969), a short paper that gives no indication of how he discovered his formulas. Several matrix problems that may be reduced to multiplication and therefore have $O(n^{2.81})$ solutions are described in Aho, Hopcroft, and Ullman (1974).

Versions of the fast Fourier transform are presented in Cooley and Tukey (1965) and in Aho, Hopcroft, and Ullman (1974). Gentleman and Sande (1966) discuss applications of the FFT and some special implementation situations, e.g., when the data do not all fit in core. Brigham (1974) is an entire book on the FFT. Aho, Hopcroft, and Ullman (1974) present an application of the FFT to integer multiplication. (The string of digits $d_n d_{n-1} \ldots d_1 d_0$ representing an integer in base b is a polynomial $\sum_{i=0}^{n} d_i b^i$.) There are many other references on the FFT since it is widely used.

Transitive Closure, Boolean Matrices, and Equivalence Relations

6.1 *The Transitive Closure of a Binary Relation*

Let S be a finite set with elements $s_1, s_2, \ldots, s_n$. A *binary relation* on S is a subset, say A, of $S \times S$. If $(s_i, s_j) \in A$ we say that s_i is A-related to s_j and use the notation $s_i A s_j$. The relation A can be represented by an $n \times n$ matrix with entries

$$a_{ij} = \begin{cases} 1 & \text{if } s_i A s_j \\ 0 & \text{otherwise}. \end{cases}$$

The adjacency relation on the set of vertices of a graph or digraph, used extensively in Chapter 3, is an important example of a relation. Other common examples of relations are equivalence relations and partial orders. We will use the same (capital) letter to denote a relation and its matrix representation (which assumes a particular ordering on the elements of the underlying set), and the corresponding lowercase letters for the matrix entries. Unless otherwise stated, we assume the set in question is $S = \{s_1, \ldots, s_n\}$.

The *transpose* of a binary relation A on S is the binary relation A^T such that (s_i, s_j) is in A^T if and only if (s_j, s_i) is in A. The matrix for A^T is the transpose of the matrix for A. (Note that by these definitions the adjacency matrix defined in Chapter 3 for graphs and digraphs is the matrix for the transpose of the adjacency relation.)

A relation A on S is *transitive* if and only if for all i, j, and k between 1 and n, $s_i A s_j$ and $s_j A s_k$ implies $s_i A s_k$. Equivalence relations and partial orders are transitive relations. Usually, the adjacency relation for a graph or digraph is not. The *transitive closure* of a relation A on S is the relation R defined by $s_i R s_j$ if and only if $s_i A s_j$ or there* exist $m \geqslant 2$ and $s_{k_1}, s_{k_2}, \ldots, s_{k_m}$ in S such that $s_i = s_{k_1}, s_{k_t} A s_{k_{t+1}}$ for $1 \leqslant t \leqslant m - 1$ and $s_{k_m} = s_j$. The transitive closure of a transitive relation A is the relation A itself. The transitive closure of the transpose of the adjacency relation for a graph or digraph with vertex set V is the *reachability relation* defined by vRw if and only if there is a path (of length >0) from v to w; i.e., if w is reachable from v by following edges.

In Sections 6.1–6.4 we will study a variety of methods for finding the transitive closure of a relation. The application to graphs and digraphs is a useful one, and in fact, if we extend the definition of a digraph to allow a vertex to be adjacent to itself, any binary relation A on a finite set S may be interpreted as the digraph $G = (S, A)$. Thus, with the extended definition which we will assume throughout these sections, the problem of finding the reachability relation, or matrix, for a digraph and the problem of finding the transitive closure of a binary relation are equivalent. We will use the terminology of whichever problem seems to motivate the particular algorithm being studied. The form in which the input is given, however, will depend on how the problem arises in a particular application.

FINDING THE REACHABILITY MATRIX BY DEPTH-FIRST SEARCH

A fairly obvious way to construct R, the reachability matrix for a digraph $G = (S, A)$, is to do a depth-first search (see Section 3.4) from each vertex to find all other vertices that can be reached from it. Initially R would be the zero matrix. Visiting, or processing, a vertex s_j encountered in the depth-first search from s_i would consist of setting r_{ij} equal to 1. Thus each depth-first search fills one row of R. This may seem overly simpleminded and inefficient since during a depth-first search from, say, s_i, entries may be made in rows other than the ith row; specifically, when a vertex s_j is encountered, r_{kj} may be set to 1 for all k such that s_k is on the path from s_i to s_j. These vertices s_k may be found on the stack. How significant is this modification? Does it eliminate the need to do a depth-first search from s_k? How does it affect the amount of work done in the worst case?

Since depth-first search was illustrated by many examples in Chapter 3, we will not work out the details of an algorithm here, but just make a few comments about the amount of work done. If the adjacency list structure described in Chapter 3 is used for G and a depth-first search is done for each vertex, the worst-case running time will be $\Theta(nm)$ where $n = |S|$ and $m = |A|$. Inserting "1"'s in more than one row of R during each depth-first search, as suggested above, can improve the algorithm's behavior for many graphs, hence its average behavior, but the worst case will still be $\Theta(nm)$. See Exercise 6.2.

In Chapter 3 we defined the condensation of a digraph. Informally, the condensation is the digraph obtained by collapsing each strongly connected component to a single point; it is acyclic. We mentioned that some problems may be simplified by working with the condensation instead of the original digraph. The reachability relation for a digraph $G = (V, E)$, where $|V| = n$, may be computed as follows:

1.   Find the strong components of G using Algorithm 3.8 [in $\Theta(n + |E|)$ time]. Let G' be the condensation of G.

2.   Find the reachability relation for G'. (Any of the methods presented in this chapter may be used, but note that the depth-first search method can be simplified somewhat if the input is known to be acyclic, as it is here.)

3.   Expand the reachability relation for G' by replacing each vertex of G' by all the vertices in G that were collapsed to it $[O(n^2)$ time].

The amount of work done at step 2 and hence by this method as a whole depends on the particular digraph G. If G has several large strong components the reduction to G' may save a lot of time.

Efficient depth-first search uses linked adjacency lists. In the next section we present a fairly simple $\Theta(n^3)$ algorithm for finding the reachability matrix without using any extra data structures.

6.2   *Warshall's Algorithm*

If we interpret a binary relation A on a finite set S as a digraph, then finding elements of R, the transitive closure of the relation, corresponds to inserting edges in the digraph. In particular, for any pair of edges $\overline{s_i s_k}$ and $\overline{s_k s_j}$ inserted so far, we add the edge $\overline{s_i s_j}$. That is, we can conclude that $s_i R s_j$ if we already know that for some k, $s_i R s_k$ and $s_k R s_j$. Thus it should be very easy to convince oneself that the following algorithm computes R.

Algorithm 6.1 TRANSITIVE CLOSURE

Input: A and n, where A is an $n \times n$ matrix that represents a binary relation.

Output: R, the matrix for the transitive closure of A.

1. $R \leftarrow A$ [i.e., copy A into R]
2.   Repeat until no entry of R changes during one complete pass:
3.   **for** $i \leftarrow 1$ **to** n;
4.   **for** $j \leftarrow 1$ **to** n;
5.   **for** $k \leftarrow 1$ **to** n;
6.   **if** $r_{ik} = 1$ and $r_{kj} = 1$ **then** $r_{ij} \leftarrow 1$.

Figure 6.1 illustrates that line 2 could not be omitted. When a particular s_i and s_j are first considered, there may be no s_k that joins them. Later on in the processing, because of the insertion of other edges, we may be able to insert $\overline{s_i s_j}$, hence it must be reconsidered. The complexity of Algorithm 6.1 is proportional to n^3 times the number of repetitions of the loop in lines 3–6. Determining this number is left as an exercise since we will revise the algorithm to reduce the amount of work done.

Suppose we refer to the work done at line 6 as processing the triple (i, j, k). In Fig. 6.1, if the triple (i, k, k') were processed before (i, k', j), then none would have to

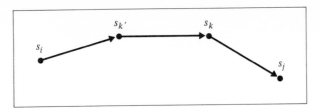

Figure 6.1
$i < k'$ and $j < k$. (k, i, j) and (k', i, j) are processed before
(k', i, k) and (k, k', j), so (k, i, j) or (k', i, j) must be
reprocessed.

be considered twice. Is there some order that eliminates the need for processing any triple more than once? Or, no matter what order we try, can we find an example in which repetition is required? We suggest that the reader try to answer these questions before proceeding.

Warshall's algorithm is simply an algorithm that processes the triples in the correct order, specifically, with k varying in the outermost loop. A proof of correctness follows the algorithm.

Algorithm 6.2 WARSHALL'S ALGORITHM FOR TRANSITIVE CLOSURE

Input: A and n, where A is an $n \times n$ matrix that represents a binary relation.

Output: R, the $n \times n$ matrix representing its transitive closure.

1. $R \leftarrow A$
2.   **for** $k \leftarrow 1$ **to** n;
3.   **for** $i \leftarrow 1$ **to** n;
4.   **for** $j \leftarrow 1$ **to** n;
5.   **if** $r_{ik} = 1$ and $r_{kj} = 1$ **then** $r_{ij} \leftarrow 1$.

Clearly the total number of triples processed is n^3. Step 1 takes $\Theta(n^2)$ time so the number of matrix entries examined and/or changed for any input is $\Theta(n^3)$.

Theorem 6.1 When Algorithm 6.2 terminates, R is the matrix representing the transitive closure of A.

Proof. For each k, let $r_{ij}^{(k)}$ be the value of r_{ij} after the middle loop (lines 3 to 5) is executed for the kth time. Let $S = \{s_1, s_2, \dots, s_n\}$ and for $0 \leq k \leq n$, define $S_k = \{s_1, \dots, s_k\}$. We claim that $r_{ij}^{(k)} = 1$ if and only if there is a path (of length > 0) from s_i to s_j using only vertices in S_k (aside from s_i and s_j, which are at the ends of the path). This characterization of $r_{ij}^{(k)}$, which clearly proves the theorem, will be established by induction on the triples (k, i, j) in lexicographical order, the same order in which the $r_{ij}^{(k)}$ are computed.

For the basis of the induction we establish the claim for $k = 0$ and all i and j; i.e., for when the loop has not yet been executed but the initialization, line 1, has been done. Then $r_{ij}^{(0)} = a_{ij} = 1$ if and only if there is an edge from s_i to s_j, which is equivalent to the claimed characterization of $r_{ij}^{(0)}$ since $S_0 = \phi$. Now suppose that $k, i, j > 0$ and the characterization holds for all triples that precede (k, i, j). Observe that when line 5 is executed for (k, i, j) the superscripts on r_{ik} and r_{kj} may be k or $k - 1$ depending on the relative order of the indexes. If $k \geqslant j$ then $r_{ik} = r_{ik}^{(k-1)}$ since r_{ik} has not yet been changed during the kth pass. The triple for $r_{ik}^{(k-1)}$, i.e., $(k - 1, i, k)$, precedes (k, i, j). Similarly, if $k \geqslant i$, then $r_{kj} = r_{kj}^{(k-1)}$ and its triple, $(k - 1, k, j)$ precedes (k, i, j). If $k < j$, then $r_{ik} = r_{ik}^{(k)}$, and its triple, (k, i, k), precedes (k, i, j); similarly for $k < i$. So the induction hypothesis may be applied to r_{ik} and r_{kj} in line 5. Therefore $r_{ij}^{(k)} = 1$ if and only if $r_{ij}^{(k-1)} = 1$ (hence, isn't changed in line 5) or both $r_{ik}^{(p)}$ and $r_{kj}^{(p')} = 1$ where $p, p' \in \{k - 1, k\}$; that is, if and only if there is a path from s_i to s_j using only vertices in S_{k-1} or if there are paths from s_i to s_k and s_k to s_j using only vertices in S_{k-1} or S_k (depending on p and p'). This finally is equivalent to there being a path from s_i to s_j using only vertices in S_k. This establishes the induction hypothesis for (k, i, j) and proves the theorem. □

If the matrices A and R are stored with one entry per bit, then Warshall's algorithm has the following particularly fast implementation using the logical **or** (or Boolean sum, or union) instruction available on most large general-purpose computers. The **or** instruction will be denoted by ∨ and is defined bitwise by: $1 \vee 0 = 0 \vee 1 = 1 \vee 1 = 1$ and $0 \vee 0 = 0$.

Algorithm 6.3 WARSHALL'S ALGORITHM FOR BIT MATRICES

Input: A and n as in Algorithm 6.2 but where A is stored as a bit matrix.

Output: R, the transitive closure of A, also as a bit matrix.

Notation: Let R_i be the ith row of R for $1 \leqslant i \leqslant n$, and let ∨ denote **or**.

1. $R \leftarrow A$
2.   **for** $k \leftarrow 1$ **to** n;
3.   **for** $i \leftarrow 1$ **to** n;
4.   **if** $r_{ik} = 1$ **then** $R_i \leftarrow R_i \vee R_k$

At most n^2 logical **or**'s are done on rows of R. However, a row may not fit in one memory word and more than one **or** instruction may be needed in line 4. (On the IBM 360 one instruction will compute the Boolean sum of two bit strings up to 256 bytes – i.e., 2048 bits – long, though the time required to execute the instruction depends on the length of the operands.) The number of **or**'s required for each row is $\lceil n/\text{word size} \rceil$ so Algorithm 6.3 does $\lceil n^3/\text{word size} \rceil$ Boolean **or** instructions in the worst case. The complexity is $\Theta(n^3)$ but the constant multiple of n^3 is small.

APPLICATION OF WARSHALL'S ALGORITHM

In Chapter 3 we studied an algorithm (Algorithm 3.3) to find a shortest path and the distance between two specified vertices in a weighted digraph. The algorithm used the linked adjacency list structure and ran in $\Theta(n^2)$ time in the worst case. Now we consider the following problem: Given a weighted digraph $G = (V, E, W)$ with $V = \{v_1, \ldots, v_n\}$, represented by the weight matrix with entries

$$w_{ij} = \begin{cases} W(\overline{v_i v_j}) & \text{if } \overline{v_i v_j} \in E \\ \infty & \text{otherwise,} \end{cases}$$

compute the $n \times n$ matrix D defined by $d_{ij} =$ the distance from v_i to v_j. Observe that this problem is more general than the problem of finding R, the reachability matrix; R can be obtained from D simply by changing all entries less than ∞ to 1's and all ∞'s to 0's.

A naive approach to computing D would be to apply Algorithm 3.3 to each pair of vertices, taking $\Theta(n^4)$ time in the worst case. A modification of Warshall's algorithm for transitive closure gives a very simple $\Theta(n^3)$ algorithm. Elements of D, like elements of R, can be computed by processing the set of triples (k, i, j) for $1 \leqslant k, i, j \leqslant n$. For D, processing the triple (k, i, j) means computing $d_{ij} \leftarrow \min\{d_{ij}, d_{ik} + d_{kj}\}$. Once again, the order in which triples are processed is critical to getting the correct result without repeated processing. The algorithm is only slightly different from Algorithm 6.2, and the proof that it is correct follows the proof of Theorem 6.1 very closely.

Algorithm 6.4 DISTANCES IN A DIGRAPH

Input: W and n, the weight matrix for a digraph with vertices $v_1, \ldots, v_n$.

Output: D, an $n \times n$ matrix such that $d_{ij} =$ the distance from v_i to v_j.

1. $D \leftarrow W$
2.   **for** $k \leftarrow 1$ **to** n;
3.   **for** $i \leftarrow 1$ **to** n;
4.   **for** $j \leftarrow 1$ **to** n;
5. $d_{ij} \leftarrow \min\{d_{ij}, d_{ik} + d_{kj}\}$

6.3   *Computing Transitive Closure by Matrix Operations*

Suppose A is the matrix for a binary relation on $S = \{s_1, \ldots, s_n\}$ and we interpret A as the adjacency relation on the digraph (S, A). Then $a_{ij} = 1$ if and only if there is a path of length one from s_i to s_j since a path of length one is an edge. Suppose we define matrices $A^{(p)}$ by

$$a_{ij}^{(p)} = \begin{cases} 1 & \text{if there is a path of length } p \\ & \text{from } s_i \text{ to } s_j \\ 0 & \text{otherwise.} \end{cases}$$

Then $A^{(0)} = I$, the identity matrix, and $A^{(1)} = A$. How can we compute $A^{(2)}$? By definition, $a_{ij}^{(2)} = 1$ if and only if there is a path of length two from s_i to s_j, hence if and only if there is a vertex s_k such that $a_{ik} = 1$ and $a_{kj} = 1$. Thus $a_{ij}^{(2)} = \bigvee_{k=1}^{n} (a_{ij} \wedge a_{kj})$, where $\vee$ is the Boolean **or** and $\wedge$ is the Boolean **and**, or product: $1 \wedge 1 = 1$ and $1 \wedge 0 = 0 \wedge 1 = 0 \wedge 0 = 0$. The formula for $a_{ij}^{(2)}$ is the formula for an entry in the product of A with itself *as a Boolean matrix*. The product $C = AB$ of $n \times n$ Boolean matrices A and B (i.e., matrices whose entries are all 1's and 0's), is by definition the Boolean matrix with entries $c_{ij} = \bigvee_{k=1}^{n} (a_{ik} \wedge b_{kj})$ for $1 \leq i, j \leq n$. For the remainder of this section all matrix products will be Boolean so we may denote $A^{(2)}$ as A^2. The *Boolean matrix sum* of A and B, $D = A \vee B$, is defined by $d_{ij} = a_{ij} \vee b_{ij}$ for $1 \leq i, j \leq n$.

We have concluded that A^2 indicates which vertices are connected by paths of length 2. It is easy to generalize and prove the following lemma by induction on p. The proof is left as an exercise.

Lemma 6.2 Let A be the adjacency matrix for a digraph with vertices $\{s_1, \ldots, s_n\}$. Denote elements of A^p, for $p \geq 0$, by $a_{ij}^{(p)}$. Then $a_{ij}^{(p)} = 1$ if and only if there is a path of length p from s_i to s_j.

The entries of R are defined by $r_{ij} = 1$ if and only if there is a path of *any* length (greater than 0) from s_i to s_j. Observe that for any p and p', the (i, j)th entry of the matrix $A^p \vee A^{p'}$ is 1 if and only if there is a path of length p or a path of length p' from s_i to s_j. Thus $R = \bigvee_{p=1}^{\infty} A^p$. ($R$ is often denoted A^+ since it is the Boolean sum of the powers of A.) This formula for R is not useful for computation since it requires computing infinitely many matrices, but the problem is easily handled by the next lemma, the proof of which is also left to the reader.

Lemma 6.3 In a digraph with n vertices, if there is a path from vertex v to vertex w, then there is a path from v to w of length at most n.

Thus, $R = \bigvee_{p=1}^{s} A^p$ for any $s \geq n$. The amount of work needed to compute R by this formula is still quite large, since s powers of the matrix A are computed. We will carry out some algebraic manipulations on the formula for R to put it in a form that suggests a more efficient computation. Some of the following properties of Boolean matrix operations will be useful. Assume that A, B, and C are $n \times n$.

$$A \vee B = B \vee A \qquad\qquad A \vee A = A$$

$$A \vee (B \vee C) = (A \vee B) \vee C \qquad A(BC) = (AB)C$$

$$A(B \vee C) = (AB) \vee (AC) \qquad (B \vee C)A = (BA) \vee (CA)$$

$$IA = AI = I, \text{ where } I \text{ is the } n \times n \text{ identity matrix.}$$

Now, let $s \geq n$. Then $R = \bigvee_{p=1}^{s} A^p = A \vee A^2 \vee \cdots \vee A^s = A(I \vee A \vee \cdots \vee A^{s-1})$.

Lemma 6.4 $I \vee A \vee A^2 \vee \cdots \vee A^{s-1} = (I \vee A)^{s-1}$ for $s \geq 1$.

Proof (by induction on s). For $s = 1$, both sides are equal to I. Assume that $s > 1$ and the equality holds for $s - 1$, i.e., $I \vee A \vee \cdots \vee A^{s-2} = (I \vee A)^{s-2}$. Then $(I \vee A)^{s-1} = (I \vee A)^{s-2}(I \vee A) = (I \vee A \vee \cdots \vee A^{s-2})(I \vee A) = I \vee A \vee A \vee A^2 \vee A^2 \vee \cdots \vee A^{s-2} \vee A^{s-2} \vee A^{s-1} = I \vee A \vee A^2 \vee \cdots \vee A^{s-1}$. $\square$

Theorem 6.5 Let A be an $n \times n$ Boolean matrix representing a binary relation. Then R, the matrix for the transitive closure of A, is $A(I \vee A)^{s-1}$ for any $s \geqslant n$.

How much work is needed to compute R using the formula of Theorem 6.5? Computing $I \vee A$ requires inserting 1's in the diagonal of A, $\Theta(n)$ operations. Since s may be any integer at least as large as n, we choose s so that $s - 1$ is a power of 2, in particular $2^{\lceil \log(n-1) \rceil}$. Then $(I \vee A)^{s-1}$ may be computed by doing $\lceil \log(n-1) \rceil$ matrix multiplications, and R is computed by a total of $\lceil \log(n-1) \rceil + 1$ matrix multiplications. Boolean matrix products may be computed as indicated by the definition in $\Theta(n^3)$ time, or in $\Theta(n^{2.81})$ time by using Strassen's algorithm (Section 5.3) and then replacing all positive entries by 1's. Computing R using these methods would then take $\Theta(n^3 \log n)$ or $\Theta(n^{2.81} \log n)$ time, respectively. Multiplying Boolean matrices is a more specialized problem than multiplying ordinary matrices with real entries, and it is worth seeking specialized algorithms to do the job faster. In the next section we develop a fast Boolean matrix multiplication algorithm for bit matrices.

6.4   *Multiplying Bit Matrices — Kronrod's Algorithm*

Let A and B be $n \times n$ Boolean matrices stored with one entry per bit. Using the logical **or** instruction, their product $C = AB$ can be computed as follows, where C_i and B_k represent the ith row of C and kth row of B, respectively, and initially all entries of C are 0:

> **for** $i \leftarrow 1$ **to** n;
>
> **for** $k \leftarrow 1$ **to** n;
>
> **if** $a_{ik} = 1$ **then** $C_i \leftarrow C_i \vee B_k$

(Compare this to Algorithm 6.3; notice how similar they are, although they compute different things.) The terminology of set union is often used for the logical **or** operation; if we view, say, A_i (the ith row of A) as the set $\{k: a_{ik} = 1\}$, and similarly for rows of B and C, then $C_i = \cup_{k \in A_i} B_k$. (All sets involved are subsets of $\{1, 2, \ldots, n\}$.) The algorithm above does at most n^2 row unions (each of which may require several logical **or** instructions). We will derive an algorithm that does fewer row unions. (The algorithm presented below is often referred to as the Four Russians' Algorithm, though it is apparently the work of M.A. Kronrod, one of the four.)

KRONROD'S ALGORITHM

Certain groups of rows of B may appear in the unions for several different rows of C. For example, suppose A is as shown in Fig. 6.2. Then $B_1 \cup B_3 \cup B_4$ is contained in rows 1, 3, and 7 of the product, and nine unions are done where three would suffice. How can some or all of the duplicated work be reduced? The approach that suggests

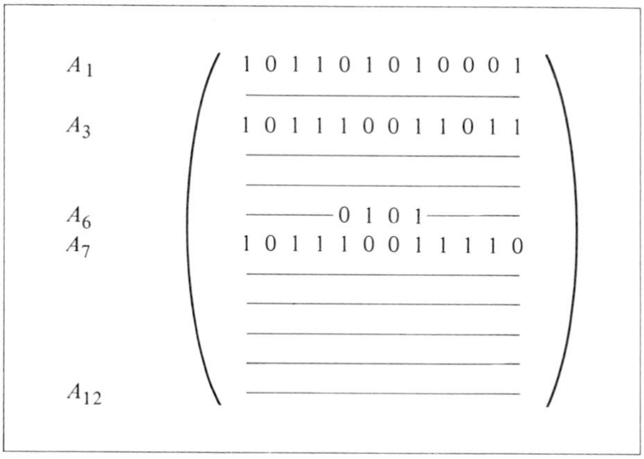

$$
\begin{array}{c}
A_1 \\[6pt]
A_3 \\[18pt]
 \\[12pt]
A_6 \\
A_7 \\[40pt]
A_{12}
\end{array}
\left(
\begin{array}{c}
1\ 0\ 1\ 1\ 0\ 1\ 0\ 1\ 0\ 0\ 0\ 1 \\[10pt]
1\ 0\ 1\ 1\ 1\ 0\ 0\ 1\ 1\ 0\ 1\ 1 \\[10pt]
\hline
\\[-6pt]
\text{—————}\ 0\ 1\ 0\ 1\ \text{——————} \\
1\ 0\ 1\ 1\ 1\ 0\ 0\ 1\ 1\ 1\ 1\ 0 \\[10pt]
\text{——————————————} \\
\text{——————————————} \\
\text{——————————————} \\[10pt]
\text{——————————————}
\end{array}
\right)
$$

Figure 6.2

itself is to first compute a lot of unions of small numbers of rows of B (like $B_1 \cup B_3 \cup B_4$), and then to combine these unions appropriately to get the rows of the product. Several questions about this approach should come to mind immediately: How many and which rows of B should be combined in the first step? How can these unions be stored so that they can be accessed efficiently during the second step? How much additional storage is needed? Will any time really be saved in the worst case? If so, how much?

The answers to most of the questions depend on the answer to the first. We adopt a straightforward strategy: divide up the rows of B into several groups of t rows each and compute all possible unions within each group. We will ignore all implementation details until we see if, with an appropriate choice for t, this strategy can produce an algorithm that does fewer than n^2 row unions. The rows of B are grouped as follows:

Group 1: $\qquad B_1, \ldots, B_t$

Group 2: $\qquad B_{t+1}, \ldots, B_{2t}$

$\qquad \vdots$

Group $\left\lceil \dfrac{n}{t} \right\rceil$: $\qquad B_p, \ldots, B_n$, where $p = \left(\left\lceil \dfrac{n}{t} \right\rceil - 1 \right) \cdot t + 1$.

Suppose, for example, that the matrix A in Fig. 6.2 is to multiply a 12×12 matrix B and let $t = 4$. Unions of all combinations of rows B_1, B_2, B_3, and B_4 would be computed once. If done in the right order (first all combinations of two rows, then three, and lastly all four), all the unions can be obtained by doing 11 row union operations. The same would be done for the groups $B_5, \ldots, B_8$ and $B_9, \ldots, B_{12}$. Then to get the first row of AB, only two more row union operations would be needed to compute $(B_1 \cup B_3 \cup B_4) \cup (B_6 \cup B_8) \cup (B_{12})$. $B_1 \cup B_3 \cup B_4$ is used again in the third and seventh rows and $B_6 \cup B_8$ is used again in the sixth row of the product.

We make a rough estimate of the total number of unions done as a function of t, and then see if we can choose a value for t that gives a total lower than n^2. For each group of rows (except perhaps the last) there are 2^t sets of rows to be combined. No unions are needed to compute the empty set or the sets consisting of just one row. Since each union of rows within a group may be computed by combining sets already computed with one more row, a total of $2^t - 1 - t$ union operations are done for each group. There are $\lceil n/t \rceil$ groups, so roughly $\lceil n/t \rceil (2^t - 1 - t)$ unions are done in the first phase of the proposed algorithm. Now any desired union of rows from B can be obtained by computing the union of at most one combination for each of the groups. Therefore, computing each row of the product matrix requires at most $\lceil n/t \rceil - 1$ additional unions, or at most $n(\lceil n/t \rceil - 1)$ additional unions for all n rows. The total number of unions done by this method is at most $\lceil n/t \rceil (2^t - 1 - t) + n(\lceil n/t \rceil - 1)$. To make our work a bit easier we approximate and consider only the high-order terms: $[(n 2^t)/t] + (n^2/t)$. If $t = 1$ or $t = n$, this expression is $\Theta(n^2)$ or $\Theta(2^n)$, respectively. Suppose that we try to minimize $[(n 2^t)/t] + (n^2/t)$ under the assumption that the first term is of higher order than the second. We would want to make t as small as possible but if $t < \log n$, the first term would no longer dominate. Similarly, if we assume that the second term is of higher order we would want t to be as large as possible, but it cannot be larger than $\log n$. This by no means rigorous argument suggests that we try $t = \log n$, or more precisely $\lceil \log n \rceil$ since t is an integer. The number of unions done for $t = \lceil \log n \rceil$ is $2n^2 / \lceil \log n \rceil$, which is of lower order than n^2. Thus this approach is worth pursuing with $t = \lceil \log n \rceil$. We will now work out some of the implementation details and determine how much extra space is needed.

For each group of rows of B there are 2^t, or $2^{\lceil \log n \rceil}$, sets to be stored. They are stored in the array UNIONS according to the following scheme. Interpret the indexes for UNIONS as t-bit binary numbers $b_1 b_2 \cdots b_t$. The bits of an index i indicate which rows of B are included in the union stored in UNIONS(i); B_j is included if and only if bit b_j in i is one. Thus the first group of unions is stored as follows:

i	Contents of UNIONS(i)
$00 \ldots 00$	$\emptyset$
$00 \ldots 01$	B_t
$00 \ldots 10$	B_{t-1}
$00 \ldots 11$	$B_{t-1} \cup B_t$
$\vdots$	$\vdots$
$11 \ldots 11$	$B_1 \cup B_2 \cup \cdots \cup B_t$

Exactly 2^t cells (each the size of one row) are used to store the unions for the first group of rows. We may suppose for now that the unions for the other groups are stored in successive blocks of cells, offset by the appropriate multiple of 2^t from the beginning of the array; later we will show how to make do with only 2^t (roughly n) cells, instead of using $2^t \lceil n/t \rceil$, or roughly $n^2/\log n$.

The storage setup described was devised to make it easy to find the unions needed for a given row of the product. Recall that the ith row of the product is $\cup_{k \in A_i} B_k$. Suppose we break up each row of A into segments of t entries each and let A_{ij} be the jth segment of t entries in the ith row. (See Fig. 6.3.) Interpreted as a binary number, A_{ij} is the correct index [minus the appropriate multiple of 2^t, in particular $(j-1)2^t$] in the array UNIONS for the union of rows of B from the jth group. At this point, the algorithm we have developed looks like this:

1. Compute and store in UNIONS unions of all combinations of rows of B within each group of $t = \lceil \log n \rceil$ successive rows.
2. **for** $i \leftarrow 1$ **to** n; [i indexes rows of A and C.]
3. **for** $j \leftarrow 1$ **to** $\lceil n/t \rceil$; [j indexes groups of rows of B.]
4. $\qquad C_i \leftarrow C_i \cup \text{UNIONS}\left((j-1)2^t + A_{ij}\right)$

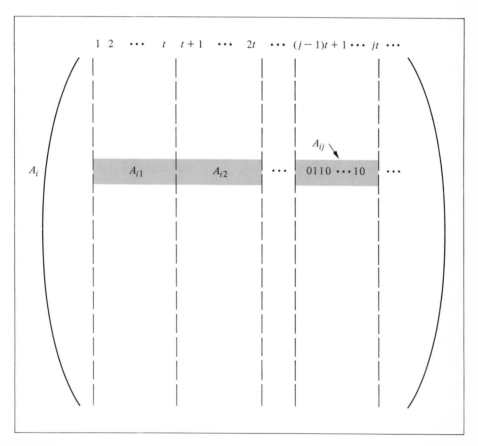

Figure 6.3

The amount of space used to store the unions can be cut down merely by changing the order in which the work is done. In its present form the algorithm computes one complete row of C before going on to the next, so all groups of unions must be available. If, instead, it works with one group at a time, selecting the union needed from that group for each row of C, succeeding groups of unions could use the same memory locations.

The last two details to work out are an efficient scheme for computing the unions within each group and a way of handling the case when the last group has less than t rows. We leave the latter problem to the reader. The former is easily taken care of in the final form of the algorithm.

Algorithm 6.5 BIT MATRIX MULTIPLICATION – KRONROD'S ALGORITHM

Input: A, B, and n, where A and B are $n \times n$ bit matrices.

Output: C, the Boolean matrix product.

Comment: B_i and C_i are the ith rows of B and C. A_{ij} denotes the jth segment of t bits in the ith row of A. As written, the algorithm assumes t divides n.

1. $t \leftarrow \lceil \log n \rceil$; UNIONS(0) $\leftarrow 0$
2.   **for** $j \leftarrow 1$ **to** $\lceil n/t \rceil$ **do**
        [Compute all unions within the jth group of rows of B.]
3.   **for** $k \leftarrow 0$ **to** $t - 1$;
4.   **for** $i \leftarrow 0$ **to** $2^k - 1$;
5.   UNIONS $(i + 2^k) \leftarrow$ UNIONS$(i) \cup B_k$
        [Select the appropriate union for each row of C.]
6.   **for** $i \leftarrow 1$ **to** n,
7. $C_i \leftarrow C_i \cup$ UNIONS (A_{ij})
    end

ANALYSIS

Note that $2^t - 1$ union operations are done to get all unions within a group (lines 3–5 because single rows of B are computed as $\phi \cup B_k$. If the loop at line 4 were started at $i = 1$, the extra t unions could be eliminated but the sets $B_1, \ldots, B_t$ must be put in the UNIONS array anyway; it is not likely that time would be saved by having another loop to insert them without doing unions. Algorithm 6.5 does a total of $(n/t)(2^t - 1 + n)$, or $\Theta(n^2/\log n)$ row unions. Each of these requires $\lceil n/\text{word size} \rceil$ Boolean **or** instructions so the running time is $\Theta(n^3/\log n)$, but is a fairly small multiple of $n^3/\log n$. The running time does not depend on the particular input; the same operations are done for all inputs of size n.

The formula derived in Section 6.3 for the matrix of the transitive closure of a relation requires approximately $\log n$ matrix multiplications. Thus Kronrod's algorithm yields a good $\Theta(n^3)$ algorithm for transitive closures.

Note that both Warshall's algorithm for transitive closure (Section 6.2) and Kronrod's Boolean matrix multiplication algorithm save time or space by doing their computations in a particular order. In both cases the natural, or more usual, order in which one would think of doing the work is less efficient.

LOWER BOUND

Is Kronrod's algorithm optimal? Since the various algorithms for Boolean matrix multiplication assume different representations for the matrices (bit matrices versus one entry per word) and do different kinds of operations (for example, Boolean operations on words, arithmetic operations if Strassen's method is used, or row unions as in Kronrod's algorithm), it is hard to establish a lower bound for a class of algorithms that includes them all. We will restrict our attention to the class of algorithms that compute rows of the product by forming unions of rows of the second factor matrix, and we will show that, within this class, Kronrod's algorithm is nearly optimal for large matrices; the number of unions done by an optimal algorithm would also be $\Theta(n^2/\log n)$. The result we prove is really not a strong one; it says that the number of row unions required is at least a very tiny bit less than $\frac{1}{4}(n^2/\log n)$ for *sufficiently large* n. The n may have to be much larger than the dimension of any matrices to be multiplied in practice. One of the reasons for including the proof of the theorem is that it illustrates a "counting argument," a useful approach for establishing lower bounds which involves counting all possible algorithms (ignoring differences not relevant to the sequence of basic operations — in this case, row unions done by the algorithms).

The phrase "for sufficiently large n" means for all n larger than some particular integer. A synonymous phrase is "almost everywhere," abbreviated a.e.

Theorem 6.6 Let ϵ be a real number between 0 and 1. Then, for sufficiently large n, any algorithm that does Boolean matrix multiplication using row unions must do at least $n^2/(4 + \epsilon)[\log n]$ union operations to multiply $n \times n$ matrices in the worst case.

Proof. Let P be an algorithm that computes $C = AB$ by forming unions of rows of B (and possibly copying rows) and can do no other operations on B. For a particular input, A and B, we can make an indexed list of the union operations done by P, denoting such an operation by UNION(r, s), where r and s may be a row of B or the result of a previous UNION specified by its index in the list. For example, an algorithm might carry out the following sequence of UNION operations if

$$
A = \begin{pmatrix} 1 & 1 & 0 & 1 \\ 1 & 0 & 1 & 1 \\ 1 & 1 & 1 & 1 \\ 0 & 0 & 0 & 1 \end{pmatrix}.
$$

1. UNION (row 1, row 4)

2. UNION (1, row 2)

3. UNION (1, row 3)

4. UNION (2, 3)

The sequence of UNION's done is not sufficient to describe the result produced by the algorithm; we must know which unions computed in the sequence are to be rows of the product, and which rows in the product they are. Suppose that A and B are $n \times n$. Then this information can be provided by an n-vector $V = \{j_1, \ldots, j_n\}$, where $-n \leqslant j_i \leqslant F(n)$ and j_i describes the ith row of the product as follows: If $j_i > 0$, the ith row is the result of the j_ith UNION operation; if $j_i = 0$, the ith row is all zeroes (the empty set); and if $j_i < 0$, the ith row is the $|j_i|$th row of B. The vector V for the example above is $(2, 3, 4, -4)$.

Suppose P is at least as efficient as Kronrod's algorithm for sufficiently large n; i.e., for all but perhaps finitely many values of n, P does at most $2n^2/\log n$ row unions. Let $F(n)$ be the number of unions done by P to multiply an arbitrary $n \times n$ matrix A and the identity matrix I_n in the worst case. The number of unions done by P in the worst case for all inputs is at least $F(n)$, and by the choice of P, any lower bound derived for $F(n)$ is a lower bound for any algorithm in the class under consideration. We will show that for $0 < \epsilon < 1$, $F(n) \geqslant n^2/[(4 + \epsilon) \log n]$ a.e.

Let S_n be the set of all valid sequences of $F(n)$ union operations. (A sequence is valid if for each i, the ith operation refers to rows of B between 1 and n and/or to the results of operations with indexes between 1 and $i - 1$.) Let V_n be the set of all n vectors with integer entries between $-n$ and $F(n)$. The operations done by P and its output for a given input A are described by an element of $S_n \times V_n$. If P does fewer than $F(n)$ unions for a particular A, S_n contains a sequence that does the work of P and is then padded out to length $F(n)$ with repetitions of, say, UNION(1, 1). We will derive an upper bound and a lower bound on $|S_n \times V_n|$ and use the resulting inequality to get a lower bound for $F(n)$.

Since each UNION has two operands, each of which is a row of B or an index between 1 and $F(n)$, there are $(n + F(n))^2$ choices for each UNION operation. Therefore $|S_n| \leqslant (n + F(n))^{2F(n)}$. $|V| = (n + 1 + F(n))^n$, so $|S_n \times V_n| \leqslant (n + 1 + F(n))^{2F(n)+n}$. To get a lower bound on $|S_n \times V_n|$, observe that $S_n \times V_n$ contains a distinct element for each $n \times n$ matrix A since $A_1 I_n \neq A_2 I_n$ if $A_1 \neq A_2$. Thus $|S_n \times V_n| \geqslant 2^{n^2}$ since there are 2^{n^2} $n \times n$ Boolean matrices. So

$$2^{n^2} \leqslant |S_n \times V_n| \leqslant (n + 1 + F(n))^{2F(n)+n}$$

or

$$n^2 \leqslant (2F(n) + n) \log (n + 1 + F(n)) \quad \text{for all } n > 0. \tag{6.1}$$

We observe that $F(n) \geqslant n^{3/2}$ a.e., for if not, Eq. (6.1) would imply that

$$n^2 \leqslant (2n^{3/2} + n) \log (n + 1 + n^{3/2}) \leqslant 3n^{3/2} \log (2n^{3/2}) \leqslant \tfrac{9}{2} n^{3/2} \log (\sqrt[3]{4n})$$

for infinitely many n, i.e., that $n^2 = O(n^{3/2} \log n)$, and this is not true. Since $F(n) \geqslant n^{3/2}$, for any $\epsilon' > 0$, $2F(n) + n < (2 + \epsilon')F(n)$ a.e. Also, $F(n) \leqslant 2n^2/\log n$ (by choice of P), so $n + 1 + F(n) < 2n^2$ a.e. Substituting these inequalities in Eq. (6.1) gives $n^2 \leqslant (2 + \epsilon') F(n) \log (2n^2) = (2 + \epsilon') F(n) (1 + 2 \log n)$ a.e. For any $\epsilon'' > 0$, $1 + 2 \log n \leqslant (2 + \epsilon'') \log n$ a.e. so $n^2 \leqslant (2 + \epsilon') (2 + \epsilon'') F(n) \log n$ a.e. Suppose that $0 < \epsilon < 1$ is given. Let $\epsilon' = \epsilon'' = \epsilon/6$. Then the inequality above becomes

$$n^2 \leqslant (4 + \epsilon) F(n) \log n, \quad \text{or} \quad \frac{n^2}{(4 + \epsilon) \log n} \leqslant F(n) \text{ a.e.} \qquad \square$$

REMARKS ON TRANSITIVE CLOSURE ALGORITHMS

We have examined several approaches to the problem of finding the transitive closure of a relation; these include depth-first search, Warshall's algorithm, and the bit matrix multiplication method. All of these run in $\Theta(n^3)$ time in the worst case but they do different kinds of operations and different numbers of them, and they work on input data in different forms. It would take $O(n^2)$ time to convert the input from one form to another (matrix to bit matrix, for example) so the *order* of the complexity of the algorithms does not depend on the form of the input, but the exact complexity does. The choice of algorithm may be determined by the form in which the data are given.

Since the varied algorithms mentioned above all take $\Theta(n^3)$ time, one might ask if we can show that any algorithm for transitive closure requires at least $\Theta(n^3)$ operations of some sort. This question has already been answered in the negative. If Strassen's matrix multiplication algorithm were used with the matrix formula derived in Section 6.3, roughly $4.7n^{2.81} \log n$ arithmetic operations would be done. $4.7n^{2.81} \log n$ is of lower order than n^3; however, because of the high constant and because Strassen's algorithm is not easy to implement, it does not produce a better algorithm than the others in practice. Because the various algorithms for transitive closure do very different kinds of operations (even more so than for Boolean matrix multiplication), it would be difficult to establish a lower bound on the work done for a class of algorithms that includes all we have studied.

6.5   *Dynamic Equivalence Relations and UNION-FIND Programs*

An *equivalence relation* R on a finite set S is a binary relation on S which is reflexive, symmetric, and transitive; that is which, for all s, t, and u in S, satisfies the properties: sRs; if sRt then tRs; and if sRt and tRu, then sRu. The equivalence class of an element s in S is the subset of S that contains all elements equivalent to s. The equivalence classes form a partition of S, i.e., they are disjoint and their union is S. The symbol $\equiv$ will be used from now on to denote an equivalence relation. The problem studied in this section is to represent, modify, and answer certain questions about an equivalence relation that changes with time. The section concludes with some discussion of applications of such relations and of the techniques developed to manipulate them.

The equivalence relation is initially the equality relation. The problem is to process a sequence of instructions of the following two types, where s_i and s_j are elements of S:

1.   IS $s_i \equiv s_j$?
2.   SET $s_i \equiv s_j$ (where $s_i \equiv s_j$ is not already true).

Question (1) is answered "yes" or "no." The correct answer depends on the instructions of the second type that have been received already; the answer is yes if and only if the instruction "SET $s_i \equiv s_j$" has already appeared or $s_i \equiv s_j$ can be derived by applying the reflexive, symmetric, and transitive properties to pairs that were explicitly made equivalent by the second type of instruction. The response to the latter, i.e., the SET instructions, is to modify the data structure that represents the equivalence relation so that later instructions of the first type will be answered correctly.

Consider the following example where $S = \{1, 2, 3, 4, 5\}$. The sequence of instructions is the left-hand column. The right-hand column shows the response — either a yes or no answer, or the set of equivalence classes for the relation as defined at the time.

Equivalence classes to start: $\{1\}, \{2\}, \{3\}, \{4\}, \{5\}$

1.	IS $2 \equiv 4$?	No
2.	IS $3 \equiv 5$?	No
3.	SET $3 \equiv 5$.	$\{1\}, \{2\}, \{3, 5\}, \{4\}$
4.	SET $2 \equiv 5$.	$\{1\}, \{2, 3, 5\}, \{4\}$
5.	IS $2 \equiv 3$?	Yes
6.	SET $4 \equiv 1$.	$\{1, 4\}, \{2, 3, 5\}$
7.	IS $2 \equiv 4$?	No

IMPLEMENTATION

To compare various implementation strategies we will count the operations of various kinds done by each strategy to process a sequence of n SET and/or IS instructions. We start by examining two fairly obvious data structures for representing the relation: matrices and arrays.

A matrix representation of an equivalence relation requires $|S|^2$ cells (or roughly $|S|^2/2$ if the symmetry is used). For an IS instruction only one entry need be examined; however, a SET instruction would require copying several rows. A sequence of n SET instructions (hence, a worst-case sequence of SET's and IS's) would require at least $n\,|S|$ operations.

The amount of space used can be reduced to $|S|$ by using an array, say CLASS, where CLASS(i) is a label or name for the equivalence class containing s_i. An instruction IS $s_i \equiv s_j$? requires looking up and comparing CLASS(i) and CLASS(j). For SET $s_i \equiv s_j$, each entry is examined to see if it equals CLASS(i) and, if so, is set to CLASS(j). Again, for a sequence of n SET's (hence, for a worst-case sequence) at least $n|S|$ operations are required.

Both of the methods described have inefficient aspects — the copying in the first and the search (for elements in CLASS(i)) in the latter. The next method uses links to attempt to avoid the extra work. As we shall see, the attempt is quite successful. Also, the approach used has a number of applications to other problems besides equivalence relations.

UNION-FIND PROGRAMS

The effect of a set instruction is to form the union of two subsets of S. An IS can be answered easily if we have a way of finding out which set a given element is in. Thus we turn our attention to UNION and FIND operators and a particular data structure in which they can be easily implemented. Each equivalence class, or subset, is represented by a tree. Each root will be used as a label or identifier for its tree. The instruction $r \leftarrow \text{FIND}(v)$ finds and sets r to the root of the tree containing v. The arguments of UNION must be roots; UNION(t, u) attaches together the trees with roots t and u ($t \neq u$). To make it easy to implement FIND and UNION the links in the trees point from each node to its parent (with null pointers in the roots). The pointers may be stored in an array with $|S|$ entries.

FIND and UNION would be used as follows to implement the equivalence instructions:

IS $s_i \equiv s_j$?	SET $s_i \equiv s_j$
$t \leftarrow \text{FIND}(s_i)$	$t \leftarrow \text{FIND}(s_i)$
$u \leftarrow \text{FIND}(s_j)$	$u \leftarrow \text{FIND}(s_j)$
Is $t = u$?	UNION(t, u)

A sequence of n UNION and/or FIND operations interspersed in any order will be considered an input, or UNION-FIND program, of size n. We take the number of operations on tree links (e.g., traversal, change, comparison) as the measure of work done. The obvious implementation for FIND(t) is simply to follow pointers from the node for t until the root is reached, and for UNION(t, u) to set the link in the node for t to point to the node for u. Each UNION does one link operation, and each FIND does $\ell + 1$, where ℓ is the level of the argument of the FIND in its tree. The program in Fig. 6.4, which does $(n/2) + (n/2)[(n/2) + 1]$ link operations, demonstrates that, using these methods, the worst-case time for a UNION-FIND program is at least order n^2. It is not hard to show that no such program does more than n^2 link operations so

```
1.  UNION (1, 2)

2.  UNION (2, 3)
          .
          .
          .
n
2.  UNION (n/2, n/2 + 1)

n
2 + 1.  FIND (1)
          .
          .
          .
n.  FIND (1)
```

Figure 6.4
A UNION-FIND program with
$S = \{1, 2, \ldots, n\}$, n even.

the worst case is $\Theta(n^2)$. It is reasonable to assume that $n \leqslant |S|$ so this is no worse and is perhaps better than the methods described earlier. We will do better.

The cost of the program in Fig. 6.4 is high because the tree constructed by the UNION instructions (see Fig. 6.5a) has large depth. It could be reduced by a more careful implementation of UNION aimed at keeping the trees short. Let W-UNION (for "weighted union") be the strategy that sets the root of the tree with fewer nodes to point to the root of the other tree (and, say, sets the first argument to point to the second if the trees have the same number of nodes). (Exercise 6.12 examines the possibility of using the depth rather than the number of nodes as the "weight" of each tree.) To distinguish between the two implementations of the UNION operation we will call the first one UW-UNION, for unweighted union. For W-UNION the number of nodes in each tree may be stored in the root; it can be distinguished from a pointer by the use of the sign bit. W-UNION must compare the numbers of nodes, set a pointer, and update the size of the new tree, so the cost of a W-UNION is 3, still a small constant. Now if we go back to the program in Fig. 6.4 (call it P) to see how much work it requires using W-UNION, we find that P is no longer a valid program because the arguments of UNION in instructions 3 through $n/2$ will not all be roots of trees. We may expand P to the program P' by replacing each instruction of the form UNION(i, j) by the three instructions $t \leftarrow$ FIND(i), $u \leftarrow$ FIND(j), UNION(t, u). Then using W-UNION, P' requires only $4n - 1$ operations! Figure 6.5 shows the trees constructed for P and P' using W-UNION and UW-UNION, respectively. We cannot conclude that W-UNION allows linear time implementations in all cases; P' is not a worst-case program when W-UNION is used. The following lemma helps us to obtain an upper bound on the worst case.

Lemma 6.7 If UNION(t, u) is implemented by W-UNION — i.e., so that the tree with root u is attached as a subtree of t if and only if the number of nodes in the tree with

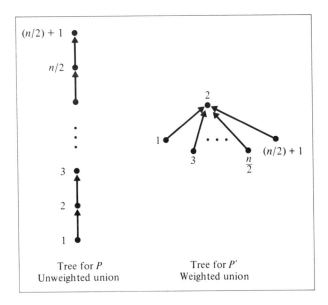

Figure 6.5
Trees obtained using unweighted union and weighted union.

root u is smaller, and the tree with root t is attached as a subtree of u otherwise — then after any sequence of UNION instructions, any tree that has k nodes will have depth at most $\lfloor \log k \rfloor$.

Proof (by induction on k). Let $k = 1$; a tree with one node has depth 0, which is $\lfloor \log 1 \rfloor$. Now suppose $k > 1$ and any tree constructed by a sequence of UNION instructions and containing k' nodes, for $k' < k$, has depth at most $\lfloor \log k' \rfloor$. Consider the tree T in Fig. 6.6 which has k nodes, depth d, and was constructed from the trees T_1 and T_2 by a UNION instruction. Suppose, as indicated in the figure, that u, the root of T_2, was set to point to t, the root of T_1. Let k_1, d_1, k_2, and d_2 be the number of nodes in and the depths of T_1 and T_2, respectively. By the induction assumption $d_1 \leqslant \lfloor \log k_1 \rfloor$ and $d_2 \leqslant \lfloor \log k_2 \rfloor$. Either $d = d_1$ (if $d_1 > d_2$) or $d = d_2 + 1$ (if $d_1 \leqslant d_2$). In the first case, $d = d_1 \leqslant \lfloor \log k_1 \rfloor \leqslant \lfloor \log (k_1 + k_2) \rfloor = \lfloor \log k \rfloor$. The second case uses the fact that, since W-UNION was used, $k_2 \leqslant k_1$. Thus $d = d_2 + 1 \leqslant \lfloor \log k_2 \rfloor + 1 = \lfloor \log 2k_2 \rfloor \leqslant \lfloor \log (k_1 + k_2) \rfloor = \lfloor \log k \rfloor$. So in both cases $d \leqslant \lfloor \log k \rfloor$.   □

Theorem 6.8 A UNION-FIND program of size n does $\Theta(n \log n)$ link operations in the worst case if the weighted union and the straightforward FIND are used.

Proof. A program that does at most n weighted union instructions can build a tree with at most $n + 1$ nodes, hence by the lemma, with depth at most $\lfloor \log (n + 1) \rfloor$, so the cost of each FIND is at most $\lfloor \log (n + 1) \rfloor + 1$. The total number of link operations is therefore less than $3n + n(\lfloor \log (n + 1) \rfloor + 1)$ which is $\Theta(n \log n + 4n) =$

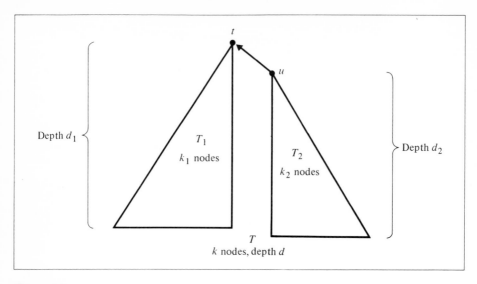

Figure 6.6
An example for the proof of Lemma 6.7.

$\Theta(n \log n)$. Showing, by example, that programs requiring $\Theta(n \log n)$ steps can be constructed is left to the reader (Exercise 6.13). $\qquad\square$

The implementation of the FIND operation can also be modified to improve the speed of a UNION-FIND program. Let C-FIND (for "collapsing-find") be the implementation which, given argument v, follows pointers from the node for v to the root, and then resets the pointers in all the nodes on the path just traversed so that they all point to the root. C-FIND is illustrated in Fig. 6.7. C-FIND does twice as much work as the straightforward FIND for a given node in a given tree, but the use of C-FIND keeps the trees very short so that overall the work will be reduced. It can be shown (see the notes and references at the end of the chapter) that using the collapsing find and the unweighted union, the worst-case running time for programs of length n is $O(n \log n)$.

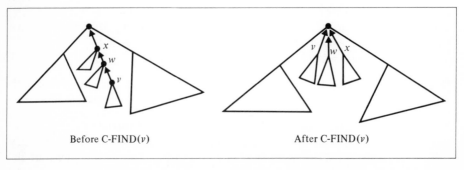

Before C-FIND(v)   After C-FIND(v)

Figure 6.7
C-FIND.

Exercises 6.14–6.18 show that there is in fact a program that requires $\Theta(n \log n)$ steps. Thus use of either the improved implementation of UNION or the improved implementation of FIND lowers the worst-case complexity of a program from $\Theta(n^2)$ to $\Theta(n \log n)$. The next step is to combine the two improvements, hoping for a further reduction.

Are C-FIND and W-UNION compatible? C-FIND changes the structure of the tree it acts on but it does not alter the number of nodes in that tree; it may, however, change the depth. Recall that it might have seemed more natural for W-UNION to compare the depths of the trees it was joining rather than the number of nodes in each since the point was to keep the trees short. It would be difficult to correctly update the depth of a tree after C-FIND modified it; the number of nodes was used specifically to make W-UNION and C-FIND compatible. We will now derive an upper bound on the number of link operations done by a UNION-FIND program using W-UNION and C-FIND. Several lemmas are needed to get the desired result: Theorem 6.12.

The *height* of a node v in a tree is the depth (i.e., height) of the subtree rooted at v. Suppose P is a UNION-FIND program of length n. Let F be the forest constructed by the sequence of UNION instructions in P assuming W-UNION is used and the FIND's are ignored. We derive a few properties of F.

Lemma 6.9 In the forest F constructed by the UNION instructions in program P using W-UNION, there are at most $n/2^{h-1}$ nodes with height h, for $h > 0$.

Proof. It follows from Lemma 6.7 that any tree with height h constructed by a sequence of W-UNION's has at least 2^h nodes. Each subtree in F (i.e., a node and all its descendants) was at one time a separate tree, so any subtree in F rooted at a node of height h has at least 2^h nodes. The set S on which the program P acts may be large, but P can affect at most $2n$ elements; the others will have height 0. For $h > 0$, since the subtrees with root at height h are disjoint, there can be at most $2n/2^h$ of them, hence at most $2n/2^h = n/2^{h-1}$ nodes of height h. $\square$

Lemma 6.10 No node has height in F greater than $\lfloor \log (n + 1) \rfloor$.

Proof. Use Lemma 6.7 and the observation that n UNION instructions can build a tree with at most $n + 1$ nodes. $\square$

Lemmas 6.9 and 6.10 describe properties of the forest F constructed by the UNION instructions of a UNION-FIND program. If the FIND instructions are executed as they occur using C-FIND, a different forest results and the heights of the various nodes will be different from their heights in F. To avoid possible confusion about which height is meant, we define the *rank* of a node to be its height in F. Thus the word "height" in Lemmas 6.9 and 6.10 may be replaced by the word "rank."

Lemma 6.11 At any time during execution of a UNION-FIND program P, the ranks of the nodes on a path from a leaf to a root of a tree form a strictly increasing sequence. When a C-FIND changes the pointer in a node, the new parent has higher rank than the old parent of that node.

Proof. Certainly in F the ranks form an increasing sequence on a path from leaf to root. If during execution of P a node v becomes a child of a node w, v must be a descendant of w in F, hence the rank of v is lower than the rank of w. If v is made a child of w by a C-FIND, then w was an ancestor of the previous parent of v; hence the second statement of the lemma follows.   □

In Theorem 6.12 we will establish an upper bound of $O(n\, G(n))$ on the running time of a UNION-FIND program using W-UNION and C-FIND, where G is a function that grows very slowly. $G(n)$ is $O(\log p(n))$ for any $p \geqslant 1$. Specifically, let $F(0) = 1$ and $F(i) = 2^{F(i-1)}$ for $i > 0$; e.g., $F(5) = 2^{2^{2^{2^2}}}$. $G(j)$ is defined as the least i such that $F(i) \geqslant j$; i.e., informally, $G(j)$ is the number of 2's that must be "piled up" to reach or exceed j. Some values of F and G are shown in Table 6.1.

Table 6.1

The functions F and G

i	0	1	2	3	4	5	6 ... 16	17 ... 65536
$F(i)$	1	2	4	16	65536	2^{65536}		
$G(i)$	0	0	1	2	2	3	3 ... 3	4 ... 4

Theorem 6.12 The number of link operations done by a UNION-FIND program of length n implemented with W-UNION and C-FIND is $O(n\, G(n))$.

Proof. The UNION instructions in the program do at most $3n$ link operations in total. The work done by the C-FIND's is counted in a tricky way. Some of the work done by each C-FIND will be assigned, or charged, to the FIND itself and some will be charged to the nodes it acts on. The nodes are grouped so that those with rank h are in group $G(h)$. (For example, nodes with rank 0 or 1 are in group 0 and nodes with ranks 6 through 16 are in group 3.) Suppose that when FIND(v) is executed, w is a node on the path from v to the root. If w is the root or if w and its parent (before being changed by the C-FIND) are in different groups, one link operation is charged to the FIND instruction itself. If w and its parent are in the same group, one is charged to v. (The total charged will be multiplied by two later because the path is traversed twice by the C-FIND.) Since the indexes of the groups containing the nodes on the path form a nondecreasing sequence, the number of operations charged to a single FIND cannot exceed the number of distinct groups, which is $1 + G(\lfloor \log(n + 1) \rfloor) \leqslant G(n + 1) \leqslant G(n) + 1$. Thus the number of operations charged to all FIND's is at most $n(G(n) + 1)$.

Next we count the operations charged to nodes. A node w is charged if it is on the path traversed by a C-FIND and it is in the same group as its parent. If w is not the child of a root, it is relinked and its new parent has higher rank than its old parent. Once a new parent of w is in a different group, w will never again be charged. Thus the total of charges to w that occur while w is not the child of a root is at most the number of ranks in its group. The ranks in group i are $F(i-1) + 1, F(i-1) + 2, \ldots,$

$F(i)$. (Look at Table 6.1.) The sum of charges for all w is at most

$$\sum_{i=0}^{G(n)} (\text{number of ranks in group } i) \cdot (\text{number of nodes in group } i). \qquad (6.2)$$

There are at most $n/2^{h-1}$ nodes of rank h so the number of nodes in group i is

$$\sum_{h=F(i-1)+1}^{F(i)} \frac{n}{2^{h-1}} = \frac{n}{2^{F(i-1)}} \sum_{j=0}^{F(i)-F(i-1)-1} \frac{1}{2^j} \leqslant \frac{n}{2^{F(i-1)}} \cdot 2 = \frac{2n}{F(i)}.$$

Thus the summation (6.2) is

$$\sum_{i=0}^{G(n)} [F(i) - F(i-1)] \cdot \frac{2n}{F(i)} = 2n \sum_{i=0}^{G(n)} \left(1 - \frac{F(i-1)}{F(i)}\right) \leqslant 2n(G(n) + 1).$$

The only operations not yet counted are those that may be charged to a node while it is the child of a root. There is at most one such operation per FIND, so there are at most n altogether. The total number of link operations done by the UNION's and FIND's is at most $3n + 2[n(G(n) + 1) + 2n(G(n) + 1) + n] = \Theta(nG(n))$. $\qquad \square$

Since $G(n)$ grows so slowly and the estimates made in the proof of the theorem are fairly loose, it is natural to wonder if we can in fact prove a stronger theorem, i.e., that the running time of UNION-FIND programs of size n implemented with W-UNION and C-FIND is $\Theta(n)$. It has been shown (see the notes and references at the end of the chapter) that this is not true; for any constant c, there are programs of size n that require more than cn operations using these techniques. It is an open question whether or not there can exist techniques that implement UNION-FIND programs in linear time. Nevertheless, as Theorem 6.12 shows, the use of C-FIND and W-UNION results in a very efficient implementation of UNION-FIND programs. We will assume this implementation when discussing the applications below. The algorithms for W-UNION and C-FIND are very easy to write out and we leave them to the reader.

EQUIVALENCE PROGRAMS

We began this section by attempting to find a good way of representing a dynamic equivalence relation so that instructions of the forms SET $s_i \equiv s_j$ and IS $s_i \equiv s_j$? could be handled efficiently. We define an equivalence program of length n to be a sequence of n such instructions interspersed in any order. Since, as we observed earlier, each SET or IS instruction can be implemented by three instructions from {UNION, FIND, equality test}, an equivalence program of length n can be implemented in $O(n\, G(n))$ time.

APPLICATION – A MINIMAL SPANNING TREE ALGORITHM

Let $G = (V, E, W)$ be a weighted graph. In Section 3.2 we studied an algorithm to find a minimal spanning tree for G. The algorithm started at an arbitrary vertex and then branched out from it by choosing edges according to a particular strategy. At any time, the edges chosen formed a tree. Here we examine an algorithm that uses a different strategy: It chooses edges in order by weight (starting with the lowest weight) and discards any edge that would form a cycle with edges already chosen. At any time the edges chosen so far will form a forest but not necessarily one tree. The correctness of the strategy is stated formally in the following theorem, the proof of which is left as an exercise.

Theorem 6.13 Let $G = (V, E, W)$ be a weighted graph and let $F \subseteq E$. If F is contained in a minimal spanning tree for G and if e is an edge of minimal weight in $E - F$ such that $F \cup \{e\}$ has no cycles, then $F \cup \{e\}$ is contained in a minimal spanning tree for G.

The algorithm starts with $F = \phi$ and continues adding edges to F until $|F| = |V| - 1$, or until it determines that G does not have a spanning tree. The only problem to be resolved is how to determine whether an edge will form a cycle with others already in F. The following lemma provides the criterion. The proof is easy and is left to the reader.

Lemma 6.14 Suppose F is a subgraph of G and F is contained in a spanning tree for G (hence F is a forest). Let $e = \overline{vw}$ be an edge of G which is not in F. There is a cycle consisting of e and edges in F if and only if v and w are in the same connected component of F.

We define a relation, $\equiv$, on the vertices in a subgraph F by $v \equiv w$ if and only if v and w are in the same connected component of F. It is easy to check that $\equiv$ is an equivalence relation. (See Fig. 6.8 for an example.) Thus, by Lemma 6.14, an edge $\overline{vw}$ is chosen if and only if it is not true that $v \equiv w$. Each time an edge is chosen, the subgraph F and the equivalence relation $\equiv$ changes; each new edge causes two connected components, or two equivalence classes, to be merged into one. Thus the minimal spanning tree algorithm presented below contains within it an equivalence program as defined above.

Algorithm 6.6 MINIMAL SPANNING TREE

Input: $G = (V, E, W)$, a weighted graph, with $|V| = n$, $|E| = m$.

Output: T, a subset of E which forms a minimal spanning tree for G, or a minimal spanning forest if G is not connected.

1.   Sort the edges in nondecreasing order by weight and put them in a list L.
2.   COUNT $\leftarrow 0$; $T \leftarrow \phi$
3.   Initialize the data structure for an equivalence program acting on elements of V.
4.   **while** COUNT $< n - 1$ **do**
5.   **if** L is empty **then exit** [G is not connected.]
6.   Let $\overline{vw}$ be the next edge in L; remove it from L.
7.   **if not** $v \equiv w$ **then do**
8. $T \leftarrow T \cup \{\overline{vw}\}$

9.   SET $v \equiv w$

10.   COUNT $\leftarrow$ COUNT + 1

                 end

    end

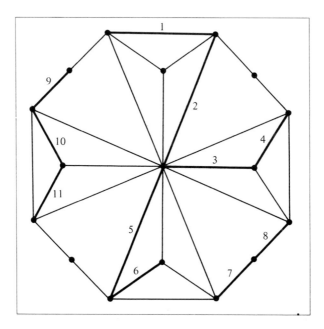

Figure 6.8
The darkened edges are in the subgraph F. The equivalence classes are $\{1, 2, 3, 4, 5, 6\}$, $\{7, 8\}$, and $\{9, 10, 11\}$.

ANALYSIS OF ALGORITHM 6.6

Line 1 can be done with $\Theta(m \log m)$ comparisons. The test in line 7 may be done as many as m times. Lines 8–10 are done at most min $\{n - 1, m\}$ times; the number of operations done in these lines therefore is bounded by a multiple of the number of operations done to execute an equivalence program of size m, i.e., it is $O(m \, G(m))$. Thus the worst-case running time of the algorithm is $\Theta(m \log m)$. Algorithm 3.2 was $\Theta(n^2)$ in the worst case. Which is better depends on the relative sizes of n and m. If the edges of G were already sorted, Algorithm 6.6 would run in $O(m \, G(m))$ time, which is quite good.

    Note that the abnormal **exit** in line 5 and the use of COUNT could be eliminated by letting the **while** loop run until L is empty. This modification would require more work for some inputs and less for others, but would not affect the worst-case complexity.

OTHER APPLICATIONS

References on all of the applications described briefly here may be found in the notes and references at the end of the chapter.

The UNION and FIND operators may be used to implement a sequence of two other types of instructions that act on the same kind of tree structures: LINK(r, v) which makes the tree rooted at r a subtree of v and DEPTH(v) which determines the current depth of v. A sequence of n such instructions can be implemented in $O(n\, G(n))$ time.

The study of equivalence programs was motivated by the problem of processing EQUIVALENCE declarations in FORTRAN and other programming languages. An EQUIVALENCE declaration indicates that two or more variables or array entries are to share the same storage locations. The problem is to correctly assign storage addresses to all variables and arrays. The declaration

$$\text{EQUIVALENCE } (A, B(3)), (B(4), C(2)), (X, Y, Z), (J(1), K), (B(1), X), (J(4), L, M)$$

	$B(1),$	$X,$ $Y,$	Z
	$B(2)$		
$A,$	$B(3),$	$C(1)$	
	$B(4),$	$C(2)$	
	$B(5),$	$C(3)$	
		$C(4)$	
		$C(5)$	
$J(1),$ K			
$J(2)$			
$J(3)$			
$J(4),$ $L,$ M			
$J(5)$			

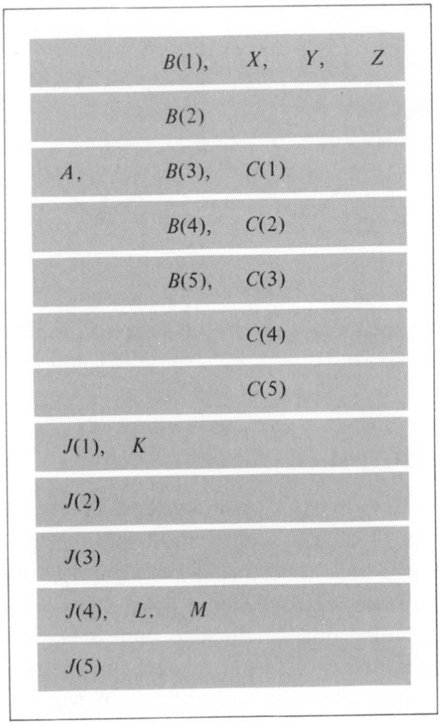

Figure 6.9
Storage arrangement for equivalence $(A,$ $B(3)), (B(4), C(2)), (X, Y, Z), (J(1), K),$ $(B(1), X), (J(4), L, M).$

indicates that A and $B(3)$ share the same location, $B(4)$ and $C(2)$ share the same location, and so forth. The complete storage layout indicated by this EQUIVALENCE statement is shown in Fig. 6.9, which assumes for simplicity that each array has five entries.

If there were no arrays, the problem of processing EQUIVALENCE declarations would be essentially the same as the problem of processing an equivalence program. The inclusion of arrays requires some extra bookkeeping and introduces the possibility of an unacceptable declaration; for example,

$$\text{EQUIVALENCE } (A(1), B(1)), (A(2), B(3))$$

could not be allowed since the elements of each array must occupy consecutive memory locations. The LINK and DEPTH instructions mentioned above can be used in the processing of EQUIVALENCE declarations.

UNION and FIND are only two of many possible operations on collections of subsets. Some others include INSERT which inserts a new member in a set, DELETE which removes an item from a set, MIN which finds the smallest item in a set, INTERSECT which produces a third set from two given, and MEMBER which indicates whether or not a specified element is in a particular set. Techniques and data structures for efficiently processing "programs" consisting of sequences of two or three types of such instructions have been studied. In some cases, the UNION and FIND techniques can be used to implement such programs of size n in $O(n\,G(n))$ time.

6.6   *Exercises*

Section 6.1: The Transitive Closure of a Binary Relation

6.1. a) Let $G = (V, E)$ be a graph (not directed) and let R be a relation on V defined by vRw if and only if $v = w$ or there exists a path from v to w. Show that R is an equivalence relation. Would R be an equivalence relation if the case $v = w$ were omitted from its definition? Why?

b) What are the equivalence classes of this relation?

c) Show that the reachability matrix R for a graph with n vertices can be constructed in $O(n^2)$ time.

6.2. a) Write an algorithm using depth-first search to construct R, the reachability matrix for a digraph, given A, the adjacency matrix. (Assume that a vertex may be adjacent to itself.) The algorithm should use the suggestion in Section 6.1 that entries of R in several rows be set during one depth-first search. Use whatever other tricks you can think of to design an efficient algorithm.

b) What is the order of the worst-case running time of your algorithm?

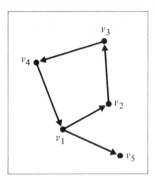

Figure 6.10

c) Test your algorithm on the digraph in Fig. 6.10. If it does not work correctly, modify it so that it does and redo (b).

d) What is the order of the running time of your algorithm on the digraph $G = (S, A)$ with $|S| = n$ and an edge $\overline{s_i s_j}$ for each $i > j$?

Section 6.2: Warshall's Algorithm

6.3. How much work is done (i.e., how many times is line 6 executed) in the worst case by Algorithm 6.1?

6.4. Construct an example of a weighted digraph on which Algorithm 6.4 would not work correctly if k were varied in the innermost loop.

6.5. Prove that Algorithm 6.4 is correct.

Section 6.3: Computing Transitive Closure by Matrix Operations

6.6. Prove Lemma 6.2.

6.7. Prove Lemma 6.3.

6.8. A *triangle* in a graph is a cycle of length 3. Outline an algorithm that uses the adjacency matrix of a graph to determine if it has a triangle. How many operations on matrix entries are done by your algorithm?

Section 6.4: Multiplying Bit Matrices – Kronrod's Method

6.9. Prove that if A and B are $n \times n$ Boolean matrices with rows interpreted as subsets of $\{1, 2, \ldots, n\}$ as described in the first paragraph of Section 6.4, then if $C = AB$, the ith row of C is $\cup_{k \in A_i} B_k$, where A_i is the ith row of A and B_k is the kth row of B.

Section 6.5: Dynamic Equivalence Relations and UNION-FIND Programs

6.10. Write algorithms for processing a sequence of SET and IS instructions using a matrix to represent the equivalence relation. Use the fact that the relation is symmetric to avoid extra work. How many matrix entries are examined or changed in the worst case when processing a list of n instructions?

6.11. Prove that a UNION-FIND program of size n does at most n^2 link operations if implemented with the unweighted union and straightforward find.

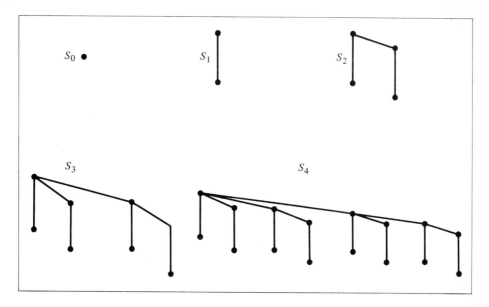

Figure 6.11
S_k trees.

6.12. The weighted union, W-UNION, uses the number of nodes in a tree as its weight. Let WD-UNION be an implementation which uses the depth of a tree as its weight and makes the tree with the smaller depth a subtree of the other.
  a) Write out an algorithm for WD-UNION.
  b) Either prove that the trees constructed for all UNION-FIND programs are the same regardless of whether W-UNION or WD-UNION is used, or exhibit a program for which they differ. (For both implementations, if the trees are of equal sizes, set the first argument to point to the second.)
  c) What is the worst-case complexity of UNION-FIND programs using the noncollapsing find and WD-UNION?

6.13. Exhibit a UNION-FIND program of size n which requires $\Theta(n \log n)$ time using the straightforward (noncollapsing) FIND and the weighted union.

6.14. S_k *trees* are defined as follows: S_0 is a tree with one node. For $k > 0$, an S_k tree is obtained from two disjoint S_{k-1} trees by attaching the root of one to the root of the other. See Fig. 6.11 for examples.
    Prove that if T is an S_k tree, T has 2^k vertices, depth k, and a unique vertex at level k. The node at level k is called the *handle* of the S_k tree.

6.15. Using the definitions and results of Exercise 6.14, prove the following characterization of an S_k tree: Let T be an S_k tree with handle v. There are disjoint trees $T_0, T_1, \ldots, T_{k-1}$, not containing v, with roots $r_0, r_1, \ldots, r_{k-1}$, respectively, such

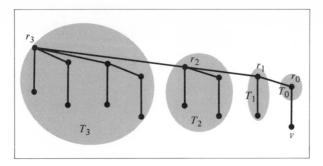

Figure 6.12
Decomposition of S_4 for Exercise 6.15.

that (1) T_i is an S_i tree, $0 \leqslant i \leqslant k - 1$, and (2) T results from attaching v to r_0, and r_i to r_{i+1}, for $0 \leqslant i < k - 1$. This decomposition of an S_4 tree is illustrated in Fig. 6.12.

6.16. Using the definitions and results of Exercise 6.14, prove the following characterization of an S_k tree: Let T be an S_k tree with root r and handle v. There are disjoint trees T_0', T_1', ..., T_{k-1}' not containing r, with roots r_0', r_1', ..., r_{k-1}', respectively, such that (1) T_i' is an S_i tree, $0 \leqslant i \leqslant k - 1$, (2) T is obtained by attaching each r_i' to r for $0 \leqslant i \leqslant k - 1$, and (3) v is the handle of T_{k-1}'. This decomposition is illustrated in Fig. 6.13 for S_4.

6.17. An *embedding* of a tree T in a tree T' is a one-one function $f: T \rightarrow T'$ (i.e., from the vertices of T to the vertices of T') such that for all v and w in T, v is the parent of w if and only if $f(v)$ is the parent of $f(w)$. An embedding f is an *initial embedding* if it maps the root of T to the root of T'; it is a *proper embedding* otherwise. Using the results of Exercises 6.14–6.16, show that if T is an S_k tree with handle v and f is a proper embedding of T in a tree U, then there is an S_k tree T' initially embedded in U', the tree that results from doing a C-FIND on $f(v)$ in U.

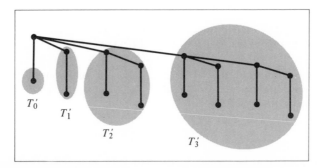

Figure 6.13
Decomposition of S_4 for Exercise 6.16.

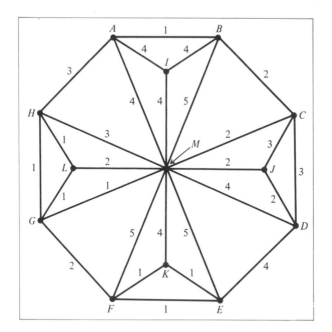

Figure 6.14
Sorted edges: *AB, EF, EK, FK, GH, GL, GM, HL, BC,*
CM, DJ, FG, JM, LM, AH, CD, CJ, HM, AI, AM, BI, DE,
DM, IM, KM, BM, EK, and *FM.*

6.18. Show that a UNION-FIND program of length n can be constructed so that if
C-FIND and UW-UNION are used to implement it, $\Theta(n \log n)$ operations are
done. (*Hint*: Read Exercises 6.14–6.17.)

6.19. Prove Theorem 6.13.

6.20. Prove Lemma 6.14.

6.21. Find the minimal spanning tree for the graph in Fig. 6.14 that would be output
by Algorithm 6.6, assuming the edges are sorted as shown.

6.22. Design an algorithm to process EQUIVALENCE declarations and assign memory
addresses to all arrays and variables in the declarations. Assume that a DIMEN-
SION statement gives the dimensions of all the arrays. Does your algorithm
detect invalid EQUIVALENCE's?

PROGRAMS

1. Write a program to compute the reachability matrix for a digraph using depth-
first search. Include an analysis of the time and space used as a function of the
input size.

2.   Write a program to multiply two bit matrices using Kronrod's algorithm (Algorithm 6.5). Allow for *n* to be larger than the number of bits per word. How much space is used?

3.   Write a program to implement UNION-FIND programs using the weighted union and collapsing find.

NOTES AND REFERENCES

Warshall's algorithm (Algorithms 6.2 and 6.3) is presented in Warshall (1962). Proofs of the correctness of Algorithm 6.2 (Theorem 6.1) and Algorithm 6.3 may be found there and in Wegner (1974). Kronrod's algorithm (Algorithm 6.5) appears in Arlazarov, Dinic, Kronrod, and Faradzev (1970) without any discussion of implementation. The proof of Theorem 6.6, the lower bound on Boolean matrix multiplication by row unions, is from Angluin (1976). This result and generalizations of Kronrod's algorithm appear in Savage (1974).

The various techniques for implementing equivalence programs and UNION-FIND programs described in Section 6.5 were developed and analyzed by many people. Galler and Fischer (1964) introduced the use of tree structures for the problem of processing EQUIVALENCE declarations. Knuth (1968) describes the equivalence problem and some suggestions for a solution (see Section 2.3.3, Exercise 1). Fischer (1972) proves that using the unweighted union and collapsing find, there are programs that do $\Theta(n \log n)$ link operations. Exercises 6.14–6.18 develop Fischer's proof. The upper bound of $O(n \log n)$ was proved by M. Paterson (unpublished). Hopcroft and Ullman (1973) proved Theorem 6.12 – i.e., that when C-FIND and W-UNION are used, a program of size *n* does $O(n G(n))$ operations. Tarjan (1975) establishes a lower bound for the worst-case behavior of C-FIND and W-UNION; he shows that it is not linear.

The application of equivalence programs to the problem of finding minimal spanning trees is attributed by Hopcroft and Ullman to M. D. McIlroy and R. Morris. Much of the material in this section plus additional applications and extensions appear in Aho, Hopcroft, and Ullman (1974).

"Hard" ($\mathcal{NP}$-Complete) Problems and Approximation Algorithms

7

7

7.1 *"Hard" Problems: Definitions, Examples, and Some Properties*

In the past six chapters we have studied quite a variety of problems and algorithms. Some of the algorithms were straightforward while others were fairly complicated and required a lot of work and creativity from the people who developed them. Whether difficult or easy to develop, virtually all of the algorithms we examined are $O(n^3)$ in complexity, where n is the appropriately defined input size, and we have been very careful to distinguish between those that are, say, $\Theta(n^2)$ in the worst case and those that are $\Theta(n \log n)$. From the point of view taken in this chapter we will accept all the algorithms studied so far as having fairly low time requirements. Here we examine a class of problems for which no reasonably fast algorithms have been developed. Many of these are optimization problems that arise frequently in applications; thus the lack of efficient algorithms is of real importance. In this section we present definitions aimed at classifying problems according to the time required to solve them, thus enabling us to distinguish between "hard" (i.e., very time-consuming) and "not-so-hard" problems. We also describe some properties of and relationships among the hard ones. Proofs for many of the important theorems presented are not included; they may be found in the references. In Sections 7.2–7.5 we will consider one approach that may be taken when hard problems must be solved quickly.

Before getting into the formal definitions and theorems, we describe several problems that will be used as examples throughout this chapter. Observe that each of these problems can be formulated so that for a given input it requires a simple yes or no answer. Several are stated as optimization problems as well.

Graph coloring A *coloring* of a graph $G = (V, E)$ is a mapping $C: V \rightarrow S$, where S is a finite set (of "colors"), such that if $\overline{vw} \in E$ then $C(v) \neq C(w)$; in other words, adjacent vertices are not assigned the same color. The chromatic number of G, denoted $\chi(G)$, is the smallest number of colors needed to color G, that is, the smallest k such that there exists a coloring C for G and $|C(V)| = k$. The problem is to produce an optimal coloring, i.e., one that uses only $\chi(G)$ colors. Alternatively, we may be given G and a positive integer k and be asked if there is a coloring of G using k colors. (If so, G is said to be k-colorable.)

The graph coloring problem is an abstraction of certain types of scheduling problems. Suppose the final exams at your university are to be scheduled during one week with three exam times each day, for a total of 15 time slots. Some courses, Calculus 1 and Physics 1 for example, must have their exams at different times because many students are in both classes. Suppose that a list of all courses is given along with a list of all groups of courses whose exams must not be at the same time. The exam scheduling problem is a graph coloring problem; the reader should be able to describe the graph involved and put an interpretation on the colors.

Job scheduling with penalties Suppose there are n jobs $J_1, \ldots, J_n$ to be executed one at a time. We are given execution times $t_1, \ldots, t_n$, deadlines $d_1, \ldots, d_n$ (measured from the starting time for the whole procedure), and penalties for missing the deadlines $p_1, \ldots, p_n$. Assume the execution times, deadlines, and penalties are all positive integers. A specific schedule for the jobs is a permutation π of $\{1, 2, \ldots, n\}$, where $J_{\pi(1)}$ is the job done first, $J_{\pi(2)}$ is the job done next, etc. The total penalty for a particular schedule is

$$P_\pi = \sum_{j=1}^{n} [\text{if } t_{\pi(1)} + \cdots + t_{\pi(j)} > d_{\pi(j)} \text{ then } p_{\pi(j)} \text{ else } 0.]$$

The problem is to find a schedule that minimizes the total penalty or, alternatively, given a nonnegative integer k, to determine if there is a schedule with $P_\pi \leq k$.

Bin packing Suppose we have an unlimited number of bins each of capacity one, and n objects with sizes $s_1, s_2, \ldots, s_n$ where $0 < s_i \leq 1$. What is the smallest number of bins into which the objects can be packed? An alternative formulation of the problem gives an integer k and asks if the objects will fit in k bins.

Applications of bin packing include packing data in computer memories (e.g., files on disk tracks, program segments into memory pages, and fields of a few bits each into memory words) and filling orders for a product (e.g., fabric or lumber) to be cut from large, standard-sized pieces.

Knapsack Given a set of n objects with sizes $s_1, \ldots, s_n$ and a knapsack with capacity C, where $s_1, \ldots, s_n$ and C are positive integers, which subset of the objects will fill the knapsack most completely? (Formally, find a 0/1 vector X that maximizes $\sum_{i=1}^{n} s_i x_i$ subject to the requirement that $\sum_{i=1}^{n} s_i x_i \leq C$.) An alternative formulation of the problem asks if there is a subset that exactly fills the knapsack.

CNF-satisfiability A logical (or Boolean) variable is a variable that may be assigned the value TRUE or FALSE. If v is a logical variable, then $\bar{v}$, the negation of v, has the value TRUE if and only if v has the value FALSE. A literal is a logical variable or the negation of a logical variable. A clause is a sequence of literals separated by the

Boolean operator ∨ **(or)**, and a logical expression in conjunctive normal form (CNF) is a sequence of clauses separated by the operator ∧ **(and)**. An example of a logical expression in CNF is $(p \vee q \vee s) \wedge (\bar{q} \vee r) \wedge (\bar{p} \vee r) \wedge (\bar{r} \vee s) \wedge (\bar{p} \vee \bar{s} \vee \bar{q})$, where p, q, r, and s are logical variables. The CNF-satisfiability problem is to determine if there is a truth assignment, i.e., a way to assign the values TRUE and FALSE, for the variables in the expression so that the expression has value TRUE.

This problem has applications in computerized theorem proving and played a central role in the development of the theoretical work to be discussed in this chapter.

Hamilton paths and Hamilton circuits   A Hamilton path (circuit) in a graph or digraph is a path (cycle) that passes through every vertex exactly once. Does a given graph or digraph have a Hamilton path (circuit)? For a related problem, the minimal tour problem, the input is a weighted graph and a minimal weighted Hamilton circuit must be found.

The usefulness and apparent simplicity of these problems may intrigue the reader; you are invited to try to devise algorithms for some of them before proceeding.

THE CLASS 𝒫

None of the algorithms devised for the problems described above are known to run in a reasonable amount of time. We will not rigorously define "reasonable," but we will define a class 𝒫 of problems that *includes* those with reasonably efficient algorithms. An algorithm is said to be polynomial-bounded if its worst-case complexity is bounded by a polynomial function of the input size, i.e., if there is a polynomial p such that for each input of size n the algorithm terminates after at most $p(n)$ steps. A problem is said to be polynomial-bounded if there is a polynomial-bounded algorithm for it. All of the problems and algorithms studied in Chapters 1–6 are polynomial-bounded. For some of the theoretical work described in this section it is convenient to consider only problems whose outputs are "yes" or "no". The ideas behind many of the definitions and theorems can be applied to the kinds of problems we have been studying – that is, ones whose outputs are matrices, spanning trees, lists of numbers, etc. – and the reader usually will not go wrong by thinking of them in this more general setting.

𝒫 is the class of problems (from among those whose output, or answer, for each input is simply "yes" or "no") that are polynomial-bounded. It may seem rather extravagant to use existence of a polynomial time bound as the criterion for defining the class of more or less reasonable problems – polynomials can be quite large. There are, however, a number of good reasons for this choice. First, while it is not true that every problem in 𝒫 has an acceptably efficient algorithm, we can certainly say that if a problem is *not* in 𝒫 it will be extremely expensive and probably impossible in practice to solve. All of the problems described at the beginning of this section are probably not in 𝒫; there are no algorithms for them that are known to be polynomial-bounded and it is believed by most researchers in the field that no such algorithms exist. Thus while the definition of 𝒫 is too broad to provide a criterion for problems with low time requirements, it provides a useful criterion – not being in 𝒫 – for problems that require too much time.

An algorithm for a complex problem may be obtained by combining several algorithms for simpler problems. The simpler algorithms may all work on the same input or some may work on the output or intermediate results of others. The complexity of the new algorithm may be bounded by addition, multiplication, and composition of the complexities of its component algorithms. Since polynomials are closed under these operations, any algorithm built from several polynomial-bounded algorithms in various ways will also be polynomial-bounded. No smaller class of functions that are useful complexity bounds has these closure properties.

There are a number of formal models of computation (classes of algorithms) used to prove rigorous theorems about the complexity of algorithms and problems. The models differ in the kinds of operations permitted, the memory resources available, and the costs assigned to different operations. A problem that requires $\Theta(f(n))$ steps on one model may require more than $\Theta(f(n))$ steps on another. A third reason for using polynomial complexity to define $\mathcal{P}$ is that for virtually all of the realistic models, if a problem is polynomial-bounded for one, then it is polynomial-bounded for the others. Thus the class $\mathcal{P}$ is invariant for a large set of often used formal models of computation.

THE CLASS $\mathcal{NP}$

Some of the sample problems described at the beginning of this section may seem easier than others and, in fact, the worst-case complexities of the algorithms that have been devised and analyzed for them do differ (they are functions like 2^n, $(n/2)^{n/2}$, $n!$, etc.), but these problems have an important and surprising property: If any one of them is in $\mathcal{P}$, i.e., if a polynomial-bounded algorithm can be devised for one of them, then all of them are in $\mathcal{P}$ and a very, very large class of problems called $\mathcal{NP}$, which includes many for which no polynomial-bounded algorithms are known, would also be contained in $\mathcal{P}$.

Many of the yes/no problems (including all of our sample problems) are phrased as existence questions: Does there exist a k-coloring of the graph G? Does there exist a truth assignment that makes a given logical expression true? For a given input, a "solution" is an object (e.g., a graph coloring or a truth assignment) that satisfies the criteria in the problem and hence justifies a yes answer (e.g., the graph coloring uses at most k colors; the truth assignment makes the CNF expression true). A "proposed solution" is simply an object of the appropriate kind – it may or may not satisfy the criteria. Loosely speaking, $\mathcal{NP}$ is the class of yes/no problems for which a given proposed solution for a given input can be checked quickly to see if it is a real solution, i.e., if it satisfies all the requirements of the problem. A proposed solution may be described by a string of symbols from some finite set (e.g., the set of characters on a keypunch machine or a computer terminal). For the formal definition of $\mathcal{NP}$ it is necessary to be able to check quickly that the string makes sense (that is, has the correct syntax) as a description of a proposed solution as well as to check that it satisfies the requirements of the problem. Thus we could think of a proposed solution as any string of characters! Suppose that we have chosen a particular set of symbols and that we have rules for describing graphs, sets, functions, etc., using these symbols. The size of a string is the number of symbols in it. We define $\mathcal{NP}$ as the class of problems for which there are polynomials p and q such that

1. There is an algorithm that determines if a given string of symbols of size s describes a solution for a given input of size n in at most $p(s + n)$ steps;

2. If there is any solution at all for an input of size n, then there is a solution of size at most $q(n)$.

Thus for an input of size n for a problem in $\mathcal{NP}$ we can give the proper answer (yes or no) if we check all strings of length at most $q(n)$. The number of steps needed to check each one is at most $p(q(n) + n)$.

All of the sample problems are in $\mathcal{NP}$, and $\mathcal{P}$ is contained in $\mathcal{NP}$. It is believed that $\mathcal{NP}$ is a much larger set than $\mathcal{P}$, but *there is not one single problem in $\mathcal{NP}$ for which it has been proved that the problem is not in $\mathcal{P}$*. There are no polynomial-bounded algorithms known for many problems in $\mathcal{NP}$, but no larger-than-polynomial lower bounds have been proved for these problems. Thus it is still not known if $\mathcal{NP}$ properly contains $\mathcal{P}$ or if $\mathcal{NP} = \mathcal{P}$.

The proposed solutions for an input may be generated and checked one at a time or simultaneously. The second alternative has the advantage of a polynomial-bounded time requirement and the disadvantage of being impossible in the real world; however, we will reconsider it later. The first alternative, while possible, is very slow since the set of strings to be checked is very large and, in particular, not polynomial-bounded. Of course there is a third alternative: Use some properties of the objects involved and some cleverness to devise an algorithm that does not have to examine all possible solutions. (When sorting, for example, we do not check each of the $n!$ permutations of the given n keys to see which one puts the keys in order.) The difficulty with the problems discussed in this chapter is that this third alternative has been fruitless; all the known algorithms either examine all possibilities, or, if they use tricks to reduce the work, the tricks are not good enough to give polynomial-bounded algorithms.

There is an alternate characterization of $\mathcal{NP}$ that encompasses the idea of simultaneous computation mentioned above; it therefore may seem unnatural or unrealistic, but it is important and is usually given as the definition of $\mathcal{NP}$. This definition uses "nondeterministic" computing devices which, while they don't exist in the real world, are very helpful for classifying problems and proving useful theorems. A nondeterministic computer is, informally speaking, one which at any step in its computation may be faced with two or more alternative courses of action and can produce copies of itself, including the contents of its memory, and continue the computation independently for each alternative. (The number of alternative computations being carried out at any time grows with the number of steps executed so it would not suffice to have several processors available.) Taking another point of view, we may think of a nondeterministic algorithm as one that arbitrarily chooses one of the possible courses of action each time it confronts several possibilities. The choice may be thought of as a guess made by the algorithm as to which choice leads to a solution. Using this interpretation of nondeterministic computation, we say that a nondeterministic algorithm solves a problem if there are *some* sequences of guesses that could be made to reach a solution. Suppose, for example, that the problem is to determine whether a graph is k-colorable. A nondeterministic algorithm may start by coloring the first vertex with an arbitrary color and then at each step color one more vertex and check that up to that point no pair of adjacent vertices has been assigned the same color. Each time the algorithm colors a new vertex it has k choices. The tree in Fig. 7.1 diagrams all the choices for a particular graph. (The notation $(c_{i_1}, \ldots, c_{i_m})$ describes an assignment of colors c_{i_j} to vertex v_j for $1 \leqslant j \leqslant m$.) Under the simultaneous-computation interpretation of nondeterministic computations, all of the colorings at one level of the tree are thought of as being considered at the same time. Under the arbitrary-guess interpretation, the algorithm carries out the computation described by one path from the root to a leaf. Either way we look at it the amount of time used by a nondeterministic algorithm in the worst case is bounded by the product of the depth of the tree describing the choices and the maximum time

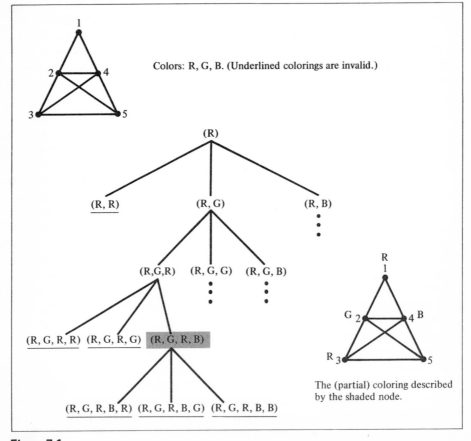

Colors: R, G, B. (Underlined colorings are invalid.)

The (partial) coloring described by the shaded node.

Figure 7.1

required for the work at each node. Each path from the root to a leaf represents the computation done to check out one possible solution. If a problem is in $\mathcal{NP}$, the tree has polynomial-bounded depth and the work done at each node is polynomial-bounded. The following statement, the usual definition of $\mathcal{NP}$, can be proved using any of several formal models of nondeterministic computation: A problem is in $\mathcal{NP}$ if and only if it can be solved by a nondeterministic algorithm in polynomial time. The name $\mathcal{NP}$ comes from "*n*ondeterministic *p*olynomial-bounded."

Observe that the tree in Fig. 7.1 also suggests a deterministic algorithm (i.e., one that does one step at a time with no guessing) for the problem: Generate and examine all the partial colorings described by the nodes in depth-first search order. Such an algorithm would do at least $(n - 1)^{n-1}$ operations in the worst case. (Let the graph be complete on n vertices and let $k = n - 1$.)

$\mathcal{NP}$-COMPLETE PROBLEMS

$\mathcal{NP}$-complete is the term used to describe problems that are the hardest ones in $\mathcal{NP}$ in the sense that if there were a polynomial-bounded algorithm for an $\mathcal{NP}$-complete problem, then there would be a polynomial-bounded algorithm for each problem in $\mathcal{NP}$. More formally, a problem P in $\mathcal{NP}$ is $\mathcal{NP}$-*complete* if given an algorithm A_P for P, we could find an algorithm for any other problem in $\mathcal{NP}$ such that there is a polynomial bound on the number of instructions it executes including subroutine, or procedure, calls to A_P, if any, but not including the work done by A_P. The following theorem follows easily from this definition.

Theorem 7.1 If any $\mathcal{NP}$-complete problem is in $\mathcal{P}$, then $\mathcal{P} = \mathcal{NP}$.

The theorem indicates, on the one hand, how valuable it would be to find a polynomial-bounded algorithm for any $\mathcal{NP}$-complete problem and, on the other hand, how unlikely it is that such an algorithm exists because there are so many problems in $\mathcal{NP}$ for which polynomial-bounded algorithms have been sought without success.

The usual technique to show that a problem is $\mathcal{NP}$-complete uses transformations of inputs for one problem into inputs for another. Suppose that A is an algorithm that converts an arbitrary input x for a problem P_1 into an input y for another problem P_2 such that the correct answer for P_1 on x is yes if and only if the correct answer for P_2 on y is yes. Then if B is an algorithm for P_2, B composed with A — i.e., B applied to the output of A — is an algorithm for P_1. If the algorithm A runs in polynomial-bounded time, we say that P_1 is *polynomial transformable* to P_2. (The term *polynomial reducible* is also used.)

The first major theorem concerning $\mathcal{NP}$-complete problems is the following.

Theorem 7.2 The CNF-satisfiability problem is $\mathcal{NP}$-complete.

The proof shows that any problem P in $\mathcal{NP}$ is polynomial transformable to CNF-satisfiability by describing an algorithm that constructs a CNF logical expression for an input x for P such that the expression, informally speaking, describes the computation of a nondeterministic algorithm for P acting on x. The expression, which is very long but constructed in time bounded by a polynomial function of the length of x, will be satisfiable if and only if the computation produces a yes answer.

Once Theorem 7.2 was proved and the importance of $\mathcal{NP}$-complete problems was recognized, many of the problems for which polynomial-bounded algorithms were being sought were shown to be $\mathcal{NP}$-complete.

Theorem 7.3 Graph coloring, the Hamilton path and circuit problems, job scheduling with penalties, bin packing, and the knapsack problem are all $\mathcal{NP}$-complete.

To prove that a problem in $\mathcal{NP}$ is $\mathcal{NP}$-complete it suffices to prove that some other $\mathcal{NP}$-complete problem is polynomial transformable to it since the polynomial transformability relation is transitive. Hence a theorem like Theorem 7.3 is proved by establishing chains of transformations beginning with the satisfiability problem. We will do two as examples.

Theorem 7.4 The directed Hamilton circuit problem is polynomial transformable to the undirected Hamilton circuit problem.

Proof. Let $G = (V, E)$ be a directed graph with n vertices. G is transformed into the undirected graph $G' = (V', E')$, where $V' = \{v^i : v \in V,\ i = 1, 2, 3\}$ and $E' = \{\overline{v^1 v^2}, \overline{v^2 v^3} : v \in V\} \cup \{\overline{v^3 w^1} : \overline{vw} \in E\}$; i.e., each vertex of G is expanded to three vertices connected by two edges, and an edge $\overline{vw}$ in E becomes an edge from the third vertex for v to the first for w. See Fig. 7.2 for an illustration. The transformation is straightforward and G' can certainly be constructed in polynomial-bounded time; it has $3n$ vertices and at most $2n + n^2$ edges.

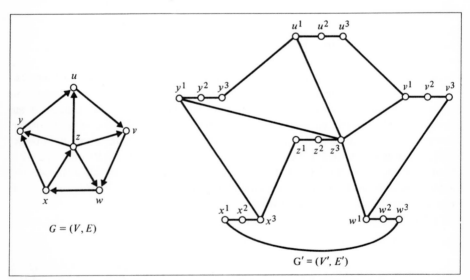

Figure 7.2
Reduction of directed Hamilton circuit problem to the undirected Hamilton circuit problem.

Now suppose G has a (directed) Hamilton circuit $v_1, v_2, \ldots, v_n$. Then v_1^1, v_1^2, v_1^3, $v_2^1, v_2^2, v_2^3, \ldots, v_n^1, v_n^2, v_n^3$ is an undirected Hamilton circuit for G'. On the other hand, if G' has an undirected Hamilton circuit, the three vertices, say v^1, v^2, and v^3, that correspond to one vertex from G must be traversed consecutively in the order v^1, v^2, v^3 or v^3, v^2, v^1 since v^2 cannot be reached from any other vertex in G'. Since the other edges in G' connect vertices with superscripts 1 and 3, if for any one triple the order of the superscripts is $1, 2, 3$, then the order is $1, 2, 3$ for all triples. Otherwise, it is $3, 2, 1$ for all triples. Since G' is undirected, we may assume its Hamilton circuit is $v_{i_1}^1, v_{i_1}^2, v_{i_1}^3, \ldots, v_{i_n}^1, v_{i_n}^2, v_{i_n}^3$. Then $v_{i_1}, v_{i_2}, \ldots, v_{i_n}$ is a directed Hamilton circuit for G. Thus G has a directed Hamilton circuit if and only if G' has an undirected Hamilton circuit. $\square$

It is of course much easier to see that the G' defined in the proof is the proper transformation to use than it is to think up the correct G' in the first place, so we make a few observations to indicate how G' was chosen. To ensure that a circuit in G' corresponds to a circuit in G, we need to simulate the directedness of the edges of G. This aim suggests giving G' two vertices, say v^1 and v^3, for each v in G with the interpretation that v^1 is used for edges in G whose head is v and v^3 is used for edges whose tail is v. Then wherever v^1, v^3 appear consecutively in a circuit for G' they can be replaced by v to get a circuit for G, and vice versa. Unfortunately, there is nothing about G' that forces v^1 and v^3 to appear consecutively in all of its circuits; thus G' could have a Hamilton circuit that does not correspond to one in G (see Exercise 7.3). The third vertex, v^2, which can only be reached from v^1 and v^3, is introduced to force the vertices that correspond to v to appear together in any circuit in G'.

Theorem 7.5 The knapsack problem is polynomial transformable to the job scheduling problem.

Proof. Let $I = (s_1, \ldots, s_n, C)$ be an input for the knapsack problem. If $\Sigma_{i=1}^n s_i < C$, then the output for I is no, and I may be transformed to any job scheduling input with a no output, e.g., $t_i = 2$, $d_i = p_i = 1$, and $k = 0$. If $\Sigma_{i=1}^n s_i \geqslant C$, then I is transformed into the following input: $t_i = p_i = s_i$ and $d_i = C$ for $1 \leqslant i \leqslant n$, and $k = \Sigma_{i=1}^n s_i - C$. Clearly the transformation takes very little time. Now suppose the knapsack input produces a yes answer; i.e., there is a subset J of $N = \{1, 2, \ldots, n\}$ such that $\Sigma_{i \in J} s_i = C$. Then let π be any permutation of N that causes all jobs with indexes in J to be done before any jobs with indexes in $N - J$. The first $|J|$ jobs are completed by their deadline since $\Sigma_{i \in J} t_i = \Sigma_{i \in J} s_i = C$ and C is the deadline for all jobs. The penalty for the remaining jobs is

$$\sum_{i=|J|+1}^n p_{\pi(i)} = \sum_{i=|J|+1}^n s_{\pi(i)} = \sum_{i \in 1} s_i - \sum_{i \in J} s_i = \sum_{i=1}^n s_i - C = k.$$

Thus the jobs can be done with total penalty $\leqslant k$.

Conversely, let π be any schedule for the jobs with total penalty $\leqslant k$. Let m be largest such that

$$\sum_{i=1}^{m} t_{\pi(i)} \leqslant C; \tag{7.1}$$

i.e., m is the number of jobs completed by the deadline C. The penalty, then, is

$$\sum_{i=m+1}^{n} p_{\pi(i)} \leqslant k = \sum_{i=1}^{n} s_i - C. \tag{7.2}$$

Since $t_i = p_i = s_i$, $1 \leqslant i \leqslant n$, we must have

$$\sum_{i=1}^{m} t_{\pi(i)} + \sum_{i=m+1}^{n} p_{\pi(i)} = \sum_{i=1}^{n} s_i,$$

and this can only happen if the inequalities in Eqs. (7.1) and (7.2) are equalities. Thus $\sum_{i=1}^{m} t_{\pi(i)} = C$ and the objects with indexes $\pi(1), \ldots, \pi(m)$ exactly fill the knapsack. $\qquad\square$

See Exercises 7.4 and 7.5 for similar problems.

WHAT MAKES A PROBLEM HARD?

If the set of inputs for an $\mathcal{NP}$-complete problem is restricted in some way, the problem may be in $\mathcal{P}$; in fact it may have a very fast solution. On the other hand, of course, even with restrictions, the problem may still be $\mathcal{NP}$-complete. For example, it is easy to test whether or not a graph (undirected) has a Hamilton circuit if the inputs are restricted to graphs with degrees at most 2. (The degree of a vertex is the number of edges incident with it.) However, even for graphs with degrees at most 3, the problem is $\mathcal{NP}$-complete. Thus it is not the possibility that vertices may have high degree that makes the problem hard. Knowing the effect on complexity of restricting the set of inputs for a problem is important because in many applications the inputs that actually occur have special properties that might allow a polynomial-bounded solution. Unfortunately, the results are discouraging; even with quite strong restrictions on the inputs many $\mathcal{NP}$-complete problems are still $\mathcal{NP}$-complete. Some of these results are described below (with proofs left for the exercises or references.) Another interesting phenomenon that is illustrated by some of the examples below is that two problems that seem to differ only slightly in their statement may differ very much in complexity; one may be in $\mathcal{P}$ while the other is $\mathcal{NP}$-complete.

The 3-CNF satisfiability problem is the CNF-satisfiability problem restricted to expressions with at most three literals per clause. It is $\mathcal{NP}$-complete. If there are at most two literals per clause, satisfiability can be checked in polynomial-bounded time. However, consider this variation of the problem: Given a CNF expression with at most two literals per clause, and given an integer k, is there a truth assignment for the variables that satisfies at least k clauses? This problem is $\mathcal{NP}$-complete.

A planar graph or digraph is one that can be drawn such that no two edges intersect. They occur in many applications, so it is well worth knowing how hard various problems are if all inputs are planar. (Determining if an arbitrary graph or digraph is planar is an important problem in itself; fortunately, it is known to be in $\mathcal{P}$ and can be solved quickly.) The directed Hamilton path problem is $\mathcal{NP}$-complete even when restricted to planar digraphs. The vertex cover problem asks, for a given graph and positive integer k, if the graph has a subset of k vertices such that each edge is incident with one of the vertices in the subset. The vertex cover problem is $\mathcal{NP}$-complete, and it is also $\mathcal{NP}$-complete even if the inputs are restricted to planar graphs. Planarity simplifies another important problem, though. A k-clique in a graph is a subgraph consisting of k mutually adjacent vertices (i.e., a complete graph on k vertices.) The problem of determining if a graph has a k-clique is $\mathcal{NP}$-complete, but for planar graphs it is in $\mathcal{P}$ because a planar graph cannot have a clique with more than four vertices. (The clique problem is also in $\mathcal{P}$ for graphs with bounded degrees.)

Although the vertex cover problem is $\mathcal{NP}$-complete, its dual, the edge cover problem — Is there a set of k edges such that each vertex is incident with one of them? — is in $\mathcal{P}$.

Determining if a graph is 2-colorable is easy; determining if it is 3-colorable is $\mathcal{NP}$-complete. It is still $\mathcal{NP}$-complete even if the graphs are planar and the maximum degree is 4. If the maximum degree is at most 2, the k-colorability problem is easy. See Exercises 7.7 and 7.8.

The feedback edge set problem is: Given a digraph $G = (V, E)$ and an integer k, is there a subset $E' \subseteq E$ such that $|E'| \leqslant k$ and every (directed) cycle in G contains an edge in E'? This problem is $\mathcal{NP}$-complete; however, the same problem for undirected graphs is in $\mathcal{P}$.

The problem of job scheduling with penalties is $\mathcal{NP}$-complete, but if the penalties are omitted and we simply asked if there is a schedule such that at most k jobs miss their deadlines, then the problem is in $\mathcal{P}$.

These examples do not yield any nice generalizations about *why* a problem is $\mathcal{NP}$-complete, but they give some indication of the effects of restricting or slightly modifying various problems. There are still a great many open questions in this field, one of which, of course, being "Is $\mathcal{P} = \mathcal{NP}$?"

Note that the more realistic formulations of problems like graph coloring and bin packing as optimization rather than yes/no problems are at least as hard to solve as the yes/no formulations — if we have an optimal coloring it is trivial to determine if a graph is k-colorable. Thus our comments about the difficulty of $\mathcal{NP}$-complete problems apply to the optimization problems associated with them, and we will refer to these optimization problems as being $\mathcal{NP}$-complete also.

7.2   *Approximation Algorithms*

The list of problems shown to be $\mathcal{NP}$-complete has grown extensively in the past few years. What can we do if we must solve one of these problems? There are several approaches to take. Even though there may not be a polynomial-bounded algorithm, there may still be significant differences in the complexities of the known algorithms;

we can try, as usual, to develop the most efficient one possible. We may concentrate on average rather than worst-case behavior and look for algorithms that are better than others by that criterion, or, more realistically, we may seek algorithms that just seem to work well for the inputs that usually occur; this choice is likely to depend on empirical tests more than on rigorous analysis. In this section we study a very different approach for solving $\mathcal{NP}$-complete optimization problems: the use of fast (or, at worst, polynomial-bounded) algorithms that are not guaranteed to give the best solution but will give one that is close to the optimal. Such algorithms are called approximation, or heuristic, algorithms. In many applications an approximate solution is good enough, especially when the time required to find the optimal solution is considered. You do not win by finding an optimal job schedule, for example, if the cost of the computer time needed to find it exceeds the worst penalty you might have paid.

The strategies, or heuristics, used by many of the approximation algorithms that have been developed are very simple and straightforward, yet for some problems they provide surprisingly good results. We will present approximation algorithms for a few $\mathcal{NP}$-complete problems.

MEASURING THE CLOSENESS OF THE APPROXIMATION

To make precise statements about the behavior of approximation algorithms (how good their results are, not how much time they take), we need several definitions. In the following paragraphs, assume that we are considering a particular optimization problem and a particular input I.

FS_I is the set of feasible solutions for I. A feasible solution is an object of the right type but not necessarily an optimal one. For example, if the problem is to find an optimal graph coloring for a given graph G, FS_G is the set of all colorings; that is, if $G = (V, E)$, $FS_G = \{C : V \to \{1, 2, \ldots, |V|\}$ such that $C(v) \neq C(w)$ if $\overline{vw} \in E\}$ (here we let the "colors" be integers.) For the bin packing problem and an input $I = \{s_1, \ldots, s_n\}$, FS_I is the set of all packings, that is, all partitions of the index set $\{1, 2, \ldots, n\}$ into disjoint subsets $T_1, \ldots, T_p$ (for some p) such that $\Sigma_{i \in T_j} s_i \leq 1$ for $1 \leq j \leq p$. The set of feasible solutions for an input to the job scheduling problem is the set of permutations of n objects, where n is the number of jobs.

The function $v_I : FS_I \to \mathbf{N}$ assigns a nonnegative integer value to each feasible solution. For graph colorings, $v_G(C) = |C(V)|$, i.e., the number of colors used. For bin packing, v_I for the partition $T_1, \ldots, T_p$ is p, the number of bins used. For job scheduling $v_I(\pi) = P_\pi$, the penalty for the schedule. It should be easy for the reader to identify the feasible solution sets and the solution value functions v_I for inputs for other optimization problems. (See Exercises 7.13 and 7.14.)

Depending on the problem, we want to find a solution that either minimizes or maximizes v_I; let "best" be "min" or "max," respectively. Then $v^*(I) = \text{best}\{v_I(x) : x \in FS_I\}$. An optimal solution is an x in FS_I such that $v_I(x) = v^*(I)$.

An *approximation algorithm* for a problem is an algorithm in $\mathcal{P}$ that, when given input I, outputs an element of FS_I. Let A be an approximation algorithm. We denote by $A(I)$ the feasible solution A chooses for I. The behavior of A on I is described by the ratio

$$r_A(I) = \frac{v(A(I))}{v^*(I)}$$

if the problem is a minimization one, and

$$r_A(I) = \frac{v^*(I)}{v(A(I))}$$

if it is maximization. In both cases, $r_A(I) \geqslant 1$. To summarize the behavior of A we define the following functions:

$$R_A(m) = \max \{r_A(I): \quad I \text{ such that } v^*(I) = m\}$$
$$S_A(n) = \max \{r_A(I): \quad I \text{ of size } n\}.$$

R_A describes the worst-case behavior of A for all inputs with a given optimal solution value, and S_A describes the worst case for a given input size. (Note that $R_A(m)$ may be ∞ for some m.)

For some problems there are approximation algorithms for which the functions R and S are arbitrarily close to 1, for others they are bounded by small constants, and for still others no algorithms guaranteed to produce reasonably close solutions are known.

7.3   The Knapsack Problem

An input for the knapsack problem is an integer vector $I = (s_1, s_2, \ldots, s_n, C)$. The problem is to find a subset T of $\{1, 2, \ldots, n\}$ to maximize $\Sigma_{i \in T} s_i$ subject to the requirement that $\Sigma_{i \in T} s_i \leqslant C$. Using the terminology and notation of Section 7.2, $FS_I = \{T:T \subseteq \{1, 2, \ldots, n\}$ and $\Sigma_{i \in T} s_i \leqslant C\}$; in other words, an approximation algorithm must output a set of objects that fit in the knapsack. The value of a feasible solution T, i.e., $v_I(T)$, is $\Sigma_{i \in T} s_i$, the amount of space filled by the objects specified by T. (The subscript I will henceforth be omitted.) An optimal solution can be found by computing $v(T)$ for each $T \subseteq \{1, 2, \ldots, n\}$ but there are 2^n such subsets. We will present a sequence of polynomial-bounded algorithms that guarantee solutions closer and closer to the optimal, though the time requirement increases with the guaranteed accuracy.

For $k \geqslant 0$ the algorithm A_k considers each subset T with at most k elements. If $\Sigma_{i \in T} s_i < C$, it goes through the remaining objects, i.e., $\{s_i : i \notin T\}$ in nonincreasing order and adds to the knapsack all that fit. The output is the set obtained that gives the largest sum. An example follows the algorithm.

Algorithm 7.1 APPROXIMATION ALGORITHM A_k FOR THE KNAPSACK PROBLEM

Input: $(s_1, s_2, \ldots, s_n, C)$, positive integers.

Output: SET, a subset of $\{1, 2, \ldots, n\}$, and MAXSUM $= \Sigma_{i \in \text{SET}} s_i$.

1. Sort the object sizes in nonincreasing order; let the resulting sequence be $s_{i_1}, s_{i_2}, \ldots, s_{i_n}$.
2. SET $\leftarrow \phi$; MAXSUM $\leftarrow 0$.
3. **for** each subset T of $\{1, 2, \ldots, n\}$ of size $\leqslant k$ **do**
4.   SUM $\leftarrow \Sigma_{i \in T} s_i$
        [Consider remaining objects.]
5.   **for** $j \leftarrow 1$ to n such that $i_j \notin T$ **do**
6.   **if** SUM $+ s_{i_j} \leqslant C$ **then do**
7.   SUM $\leftarrow$ SUM $+ s_{i_j}$
8. $T \leftarrow T \cup \{i_j\}$
            end
        end
        [See if T fills the knapsack more than the best set found so far.]
9.   **if** MAXSUM $<$ SUM **then do**
10.   MAXSUM $\leftarrow$ SUM
11.   SET $\leftarrow T$
        end
    end

Note that tests "if SUM $= C$" could be added in appropriate places to avoid consideration of extra elements and subsets once the knapsack has been filled to capacity. Such tests were omitted to keep the description of the strategy used as clear as possible.

Suppose the input for the problem is $(54, 45, 43, 29, 23, 21, 14, 1, 110)$, where $C = 110$. Table 7.1 shows the subsets considered by A_0 and A_1. The optimal solution is $\{43, 29, 32, 14, 1\}$ which fills the knapsack completely. This solution would be found by A_2.

Theorem 7.6 For $k > 0$, algorithm A_k does $O(kn^{k+1})$ operations; A_0 does $\Theta(n \log n)$. Hence $A_k \in \mathcal{P}$ for $k \geqslant 0$.

Proof. Line 1 can be done with $\Theta(n \log n)$ operations. There are $\binom{n}{j}$ subsets of $\{1, 2, \ldots, n\}$ of size j so the outer loop (instructions 4–11) is executed $\Sigma_{j=0}^k \binom{n}{j}$ times. Since $\binom{n}{j} \leqslant n^j$ and $\binom{n}{0} = 1$, $\Sigma_{j=0}^k \binom{n}{j} \leqslant kn^k + 1$. The amount of work done in one pass through the loop is clearly $O(n)$ so for all passes it is $O(kn^{k+1} + n)$. We leave it to the reader to show that the bookkeeping for systematically generating one subset of size $\leqslant k$ from the previous one can be done in $O(k) = O(1)$ time; i.e., it does not depend on n (Exercise 7.15; this is not a trivial problem). Thus the total work done is $O(\max\{n \log n, kn^{k+1} + n\})$ and the theorem follows. $\square$

Table 7.1

	Subsets of size k	Objects added at line 8	SUM
$k = 0$	ϕ	54, 45, 1	100
	MAXSUM = 100	SET = $\{54, 45, 1\}$	
$k = 1$	54	45, 1	100
	45	54, 1	100
	43	54, 1	100
	29	54, 23, 1	107
	23	54, 29, 1	107
	21	54, 29, 1	105
	14	54, 29, 1	98
	1	54, 45	100
	MAXSUM = 107	SET = $\{29, 54, 23, 1\}$	

Theorem 7.7 For $k > 0$, $R_{A_k}(m)$ and $S_{A_k}(n)$, the worst-case ratios of the optimal solution value to the value found by A_k, are at most $1 + (1/k)$ for all m and n.

Proof. Fix k and let $I = (s_1, \ldots, s_n, C)$ be a particular input with $v^*(I) = m$. Suppose an optimal solution is obtained by filling the knapsack with $s_{i_1}, s_{i_2}, \ldots, s_{i_p}$. If $p \leqslant k$, then this subset (actually the index set $\{i_1, i_2, \ldots, i_p\}$) is explicitly considered by A_k so $v(A_k(I)) = m$ and $r_{A_k}(I) = 1$. Let $p > k$, and suppose $s_{i_1}, s_{i_2}, \ldots, s_{i_p}$ are in non-increasing order. Since $m = \Sigma_{j=1}^{p} s_{i_j}$ and the jth largest of these objects can be at most one-jth of the sum, it follows that for $1 \leqslant j \leqslant p, s_{i_j} \leqslant m/j$ and, therefore, for $k + 1 \leqslant j \leqslant p$, $s_{i_j} \leqslant m/(k+1)$. At some point A_k considers the set $\{i_1, \ldots, i_k\}$ at line 3 and goes on to add more elements in lines 5–8. Let i_q be the first index in the optimal solution that is not selected by A_k. If there is no such i_q, then A_k gives an optimal solution. Otherwise the unfilled space in the knapsack is less than s_{i_q}. Thus $A_k(I) + s_{i_q} > C \geqslant m$, and, writing r for the ratio $r_{A_k}(I)$ we have

$$r = \frac{m}{v(A_k(I))} < 1 + \frac{s_{i_q}}{v(A_k(I))} \leqslant 1 + \frac{m}{(k+1)\, v(A_k(I))} = 1 + \frac{r}{(k+1)};$$

that is, $r < 1 + r/(k+1)$ which simplifies to $r < 1 + (1/k)$. Thus $R_{A_k}(m) \leqslant 1 + (1/k)$ and $S_{A_k}(n) \leqslant 1 + (1/k)$. $\square$

Corollary Given any $\epsilon > 0$, there is a polynomial-bounded algorithm A for the knapsack problem for which $R_A(m) \leqslant 1 + \epsilon$ and $S_A(n) \leqslant 1 + \epsilon$ for all m and n.

There is a more general formulation of the knapsack problem, which has more applications than the form we have used. The input consists of C and two vectors,

$S = (s_1, \ldots, s_n)$ and $P = (p_1, \ldots, p_n)$. The problem is to maximize $\Sigma_{i \in T}\, p_i$ subject to the restraint $\Sigma_{i \in T}\, s_i \leqslant C$ where T, as before, is a subset of the indexes. In an application, the elements of S would be the amount of some resource needed for each of n activities (or investments), C would be the total supply of the resource, and the elements of P would be the amount of benefit, or profit, one would get from each activity. Note that especially in applications like this where some of the data (P in this case) is only estimated or guessed at, it is not critical that an exactly optimal solution be found. We leave it as an exercise for the reader to extend the algorithms and Theorems 7.6 and 7.7 to the more general problem.

7.4   Bin Packing

Let $S = (s_1, \ldots, s_n)$ where $0 < s_i \leqslant 1$ for $1 \leqslant i \leqslant n$. The problem is to pack $s_1, \ldots, s_n$ into as few bins as possible where each bin has capacity one. An optimal solution can be found by considering all ways to partition a set of n items into n or fewer subsets, but the number of possible partitions is more than $(n/2)^{n/2}$. The approximation algorithm we present here uses a very simple strategy, called FIRST FIT; it has worst-case complexity $\Theta(n^2)$, and produces fairly good solutions. FIRST FIT means that each object in turn is placed in the first bin in which it fits. The algorithm sorts the objects first so that they are considered in order of nonincreasing size.

Algorithm 7.2 NONINCREASING FIRST FIT

Input: $S = (s_1, \ldots, s_n)$ where $0 < s_j \leqslant 1$ for $1 \leqslant j \leqslant n$.

Output: $(B_1, \ldots, B_n)$ where for $1 \leqslant j \leqslant n$, B_j is a subset of $\{1, 2, \ldots, n\}$; the elements of B_j are the indexes of the objects placed in the jth bin.

Comment: For $1 \leqslant j \leqslant n$, b_j is the amount of space in the jth bin that is already filled.

1.   Sort the elements of S in nonincreasing order. Let $s_{i_1}, s_{i_2}, \ldots, s_{i_n}$ be the result.
2.   **for** $j \leftarrow 1$ **to** n **do** $B_j \leftarrow \phi$; $b_j \leftarrow 0$ **end**
3.   **for** $t \leftarrow 1$ **to** n **do**
          [Look for a bin in which s_{i_t} fits.]
4. $j \leftarrow 1$
5.   **while** $b_j + s_{i_t} > 1$ **do** $j \leftarrow j + 1$ **end**
6. $B_j \leftarrow B_j \cup \{t\}$; $b_j \leftarrow b_j + s_{i_t}$
       end

The sort can be done in $\Theta(n \log n)$ time and j is incremented in line 5 at most $n(n-1)/2$ times. All of the other instructions are executed at most n times so the worst-case complexity is $\Theta(n^2)$ as claimed.

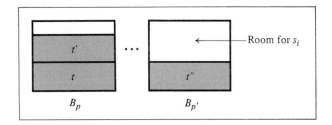

Figure 7.3

It is not difficult to construct an example where the NONINCREASING FIRST FIT algorithm (henceforth abbreviated NIFF) does not produce an optimal packing. See Exercise 7.17. Theorem 7.10, which gives upper bounds on R_{NIFF} and S_{NIFF} is established via the next two lemmas.

Lemma 7.8 Let $I = (s_1, \ldots, s_n)$ be an input for the bin packing problem and let $v^*(I)$ be the optimal (i.e., minimal) number of bins for I. All of the objects placed by NIFF in extra bins (i.e., bins with index larger than $v^*(I)$) have size at most $\frac{1}{3}$.

Proof. We may assume I is already sorted. Let i be the index of the first object placed by NIFF in bin $v^*(I) + 1$. It suffices to show $s_i \leqslant \frac{1}{3}$. We may assume $i = n$ since no object considered after s_i affects the argument. Suppose $s_i > \frac{1}{3}$. Then $s_1, \ldots, s_{i-1} > \frac{1}{3}$ so bins B_j for $1 \leqslant j \leqslant v^*(I)$ contain at most two objects each. We claim that for some $\ell \geqslant 0$ the first ℓ bins contain one object each and the remaining $v^*(I) - \ell$ bins contain two each. Otherwise there would be bins B_p and $B_{p'}$, as in Fig. 7.3, with $p < p'$ such that B_p has two objects, say t and t' with $t \geqslant t'$, and $B_{p'}$ only one, t''. Since the objects are considered in nonincreasing order, $t \geqslant t''$ and $t' \geqslant s_i$; so $1 \geqslant t + t' \geqslant t'' + s_i$ and NIFF could have put s_i in $B_{p'}$.

Thus the bins are filled by NIFF as in Fig. 7.4. Since NIFF did not put any of $s_{\ell+1}, \ldots, s_i$ in the first ℓ bins, none of them can fit; therefore, in an optimal solution the first ℓ bins will be filled the same way. Then, in an optimal solution, although they may not be arranged exactly as in Fig. 7.4, $s_{\ell+1}, \ldots, s_{i-1}$ will be in bins $B_{\ell+1}, \ldots, B_{v^*}$

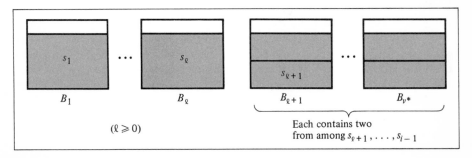

Figure 7.4

and since all are $> \frac{1}{3}$, there will be two in each of these bins and $s_i > \frac{1}{3}$ cannot fit. But an optimal solution must fit s_i in one of the first $v^*(I)$ bins; therefore the assumption that $s_i > \frac{1}{3}$ must be false. □

Lemma 7.9 For any input $I = (s_1, \ldots, s_n)$, the number of objects placed by NIFF in extra bins is at most $v^*(I) - 1$.

Proof. Since all the objects fit in $v^*(I)$ bins, $\Sigma_{i=1}^n s_i \leqslant v^*(I)$. Suppose NIFF puts $v^*(I)$ objects with sizes $t_1, \ldots, t_{v^*(I)}$ in extra bins and let b_i be the final contents of bin B_i for $1 \leqslant i \leqslant n$. If $b_i + t_i \leqslant 1$ NIFF could have put t_i in B_i, so

$$\Sigma_{i=1}^n s_i \geqslant \Sigma_{i=1}^{v^*(I)} (b_i + t_i) > v^*(I),$$

which is impossible. □

Theorem 7.10 $R_{\text{NIFF}}(m) \leqslant \frac{4}{3} + (1/3m)$. $S_{\text{NIFF}}(n) \leqslant \frac{3}{2}$ and for infinitely many n, $S_{\text{NIFF}}(n) = \frac{3}{2}$.

Proof. Let $I = (s_1, \ldots, s_n)$ be an input with $v^*(I) = m$. NIFF puts at most $m - 1$ objects, each of size at most $\frac{1}{3}$, in extra bins, so NIFF uses at most $m + \lceil (m-1)/3 \rceil$ bins. Thus

$$r_{\text{NIFF}}(I) \leqslant [m + \lceil (m-1)/3 \rceil]/m = 1 + (m + \alpha)/3m \text{ (where } -1 \leqslant \alpha \leqslant 1) \leqslant \tfrac{4}{3} + (1/3m).$$

Thus $R_{\text{NIFF}}(m) \leqslant \frac{4}{3} + (1/3m)$. For input size n, $r_{\text{NIFF}}(I)$ is largest for $m = 2$ (if $m = 1$, NIFF uses only one bin), so $S_{\text{NIFF}}(n) \leqslant \frac{4}{3} + \frac{1}{6} = \frac{3}{2}$. Construction of a sequence of examples I_n for arbitrarily large n where $r_{\text{NIFF}}(I_n) = \frac{3}{2}$ is left as an exercise (Exercise 7.17). □

A stronger result than that stated in Theorem 7.10 is known: For $\epsilon > 0$ and m large enough, $R_{\text{NIFF}}(m) \leqslant \frac{5}{4} + \epsilon$. For arbitrarily large m, there are examples that show $R_{\text{NIFF}}(m) \geqslant \frac{11}{9}$.

Another heuristic strategy that can be used for bin packing is BEST FIT: An object of size s is placed in a bin B_j which is most full among those bins in which the object fits; i.e., b_j is maximal subject to the requirement $b_j + s \leqslant 1$. If the s_i are sorted in nonincreasing order, the best fit strategy works about as well as NIFF. If the s_i are not sorted (for FIRST FIT or BEST FIT) the results can be worse but the number of bins used would still be less than twice the optimal.

There is another strategy that is even simpler than FIRST FIT and BEST FIT and gives an approximation algorithm that is faster and can be used in circumstances where the contents of all the bins cannot be stored but must be output as the packing progresses. The strategy is called NEXT FIT. The s_i are not sorted. One bin is filled at a time. Objects are put in the current bin until the next one does not fit; then a new bin is started and no more objects are packed in bins considered earlier. Thus if the sizes are, say, 0.2, 0.2, 0.7, 0.8, 0.3, 0.6, 0.3, 0.2, 0.6, they would be placed in six bins,

| 0.2, 0.2 | 0.7 | 0.8 | 0.3, 0.6 | 0.3, 0.2 | 0.6 |

although they would fit in four. Clearly the NEXT FIT strategy can be implemented in a linear time algorithm. It may seem, however, that NEXT FIT can use a lot of extra bins. In fact, its worst-case behavior is worse than NIFF, but the observation that the sum of the contents of any two consecutive bins must be greater than one allows us to conclude that $R_{\text{NEXT FIT}}(m) \leqslant 2$.

For some bin packing strategies, if the s_i are bounded by some number less than one, better (i.e., lower) bounds on the ratio of actual to optimal output can be proved.

7.5 *Graph Coloring*

For the knapsack and bin packing problems we have found approximation algorithms that give fairly good results; the behavior ratio is bounded by a small constant. A number of heuristic algorithms have been developed for the graph coloring problem but unfortunately they can produce colorings that are very far from optimal. Several of these algorithms are based on a simple strategy called SEQUENTIAL COLORING (and abbreviated henceforth as SC). Let $G = (V, E)$ where $V = \{v_1, \ldots, v_n\}$, and let the "colors" be positive integers. Beginning with the first vertex, the next vertex to be colored, say v_i, is assigned the minimum acceptable color, that is, the minimum color not already assigned to a vertex adjacent to v_i.

Algorithm 7.3 SEQUENTIAL COLORING

Input: $G = (V, E)$, a graph, where $V = \{v_1, \ldots, v_n\}$.

Output: A coloring of G.

1.   **for** $i \leftarrow$ to n **do**
2. $c \leftarrow 1$
3.   **while** there is a vertex adjacent to v_i and colored c
4.   **do** $c \leftarrow c + 1$ **end**
5.   color v_i with c
   end

Line 3 can be implemented so that it requires $O(i)$ operations so the worst-case complexity of Algorithm 7.3 is $O(n^2)$.

The behavior of the SC algorithm on a given graph depends on the ordering of the vertices. For $k \geqslant 2$, define the graphs $G_k = (V_k, E_k)$, where $V_k = \{a_i, b_i : 1 \leqslant i \leqslant k\}$ and $E_k = \{\{a_i, b_j\} : i \neq j\}$. See Fig. 7.5 for an illustration. If V is given in the order $a_1, \ldots, a_k, b_1, \ldots, b_k$, then SC will color all the a's with one color and all the b's with another, producing an optimal coloring. However, if the vertices are ordered $a_1, b_1, a_2, b_2, \ldots, a_k, b_k$, then SC needs a new color for each pair a_i, b_i, using a total of k colors. Thus $R_{\text{SC}}(2) = \infty$ and if we take $n = |V|$ as the size of a graph, $S_{\text{SC}}(n) \geqslant n/4$ for $n \geqslant 4$.

Several more complicated graph coloring algorithms based on SEQUENTIAL COLORING have additional features intended to prevent the poor behavior of SC. One such feature is to interchange two colors in the colored portion of the graph when

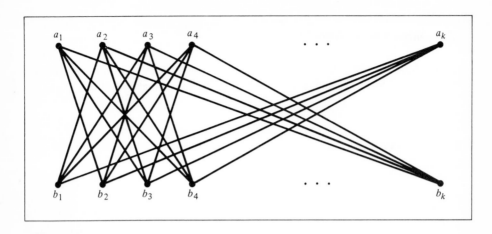

Figure 7.5
The graph G_k.

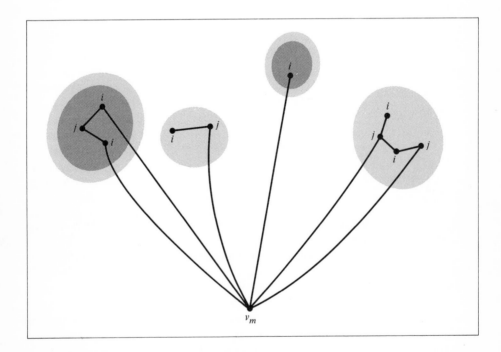

Figure 7.6
G_{ij} consists of the four connected components circled. S_i consists of the two components doubly circled.

so doing avoids the need for a new color. The expanded algorithm will be called SCI; the interchange rule is formulated as follows and is illustrated in Fig. 7.6. Suppose $v_1, \ldots, v_{p-1}$ have been colored using colors $1, 2, \ldots, c$ (where $c \geqslant 2$) and v_p is adjacent to a vertex of each color. For each pair (i, j) with $1 \leqslant i \neq j \leqslant c$ let G_{ij} be the subgraph consisting of all vertices colored i or j and all edges between these vertices. If there is a pair (i, j) such that in each connected component of G_{ij} the vertices adjacent to v_p are all of the same color, then an interchange will be done. Specifically, let S_i be the set of all vertices in connected components of G_{ij} where vertices adjacent to v_p are colored i; colors i and j are interchanged in S_i and v_p is colored with i. The algorithm then goes on to v_{p+1}.

The work needed to determine when to interchange colors and to carry out the interchange may add significantly to the time requirement of the algorithm, but the SCI strategy will produce better colorings than SC for many graphs. It will give an optimal coloring for the graphs G_k (the reader should check this), and it can be shown that SCI will yield an optimal coloring for any graph G with $v^*(G)$, i.e., $\chi(G)$, the chromatic number of G, equal to 1 or 2. However for $k \geqslant 3$ there are graphs G'_k with $3k$ vertices and $\chi(G'_k) = 3$ for which SCI uses k colors: $G'_k = \{V'_k, E'_k\}$ where $V'_k = \{a_i, b_i, c_i : 1 \leqslant i \leqslant k\}$ and $E'_k = \{\overline{a_i b_j}, \overline{a_i c_j}, \overline{b_i c_j} : i \neq j\}$. Thus $S_{SCI}(n) \geqslant n/9$ for most n, and $R_{SCI}(3) = \infty$.

The reader may observe that if the vertices in G'_k are ordered $a_1, \ldots, a_k$, $b_1, \ldots, b_k, c_1, \ldots, c_k$, then SCI produces an optimal coloring. Thus another approach to the problem of improving the basic sequential coloring strategy is to order the vertices in a special way before assigning colors. Such techniques yield significant improvements in the colorings produced for many graphs, but once again, for $k \geqslant 3$ there is a sequence of graphs G''_k for which $\chi(G''_k) = 3$ but the interchange strategy combined with an ordering strategy uses k colors.

There are at present no known polynomial-bounded graph coloring algorithms for which $S(n)$ is bounded by a constant; the best known has $S(n) = \Theta(n/\log n)$. In fact it has recently been shown that if there were an approximation algorithm for graph coloring that was guaranteed to give a coloring using at most roughly twice the optimal number of colors, then it would be possible to obtain an optimal coloring in polynomial-bounded time, and that would imply $\mathscr{P} = \mathscr{N}\mathscr{P}$. Thus getting very good but not necessarily optimal solutions for a particular $\mathscr{N}\mathscr{P}$-complete problem may itself be an $\mathscr{N}\mathscr{P}$-complete problem.

7.6   *Exercises*

Section 7.1: "Hard" Problems: Definitions, Examples, and Some Properties

7.1. Suppose algorithms A_1 and A_2 have worst-case time bound p and q, respectively. Suppose algorithm A_3 consists of applying A_2 to the output of A_1. (The input for A_3 is the input for A_1.) Give a worst-case time bound for A_3.

7.2. The knapsack problem may be stated so that $s_1, \ldots, s_n$ and C are rational numbers. Show that this version of the problem is polynomial reducible to the version in the text and vice versa.

7.3. For a directed graph $G = (V, E)$, let $G' = (V', E')$ where $V' = \{v^i : i = 1, 2, v \in V\}$ and $E' = \{\overline{v^1 v^2} : v \in V\} \cup \{\overline{v^2 w^1} : \overline{vw} \in E\}$. Show by example that there is a directed graph G such that G does not have a Hamilton circuit but G' does.

7.4. The clique problem is: Given a graph G and a positive integer k, does G have k mutually adjacent vertices? Show that CNF-satisfiability is polynomial reducible to the clique problem by using the following transformation. Suppose C_1, $C_2, \ldots, C_p$ are the clauses in a CNF expression and let the literals in the ith clause be denoted $l_{i1}, l_{i2}, \ldots, l_{iq_i}$. The expression is transformed to the graph with $V = \{(i, r) : 1 \leqslant i \leqslant p,\ 1 \leqslant r \leqslant q_i\}$, i.e., V has a vertex representing each occurrence of a literal in a clause, and $E = \{(i, r)(j, s) : i \neq r \text{ and } l_{ir} \neq \overline{l_{js}}\}$. In other words, there is an edge between two vertices representing literals in different clauses so long as it is possible for both of those literals to be assigned the value TRUE.

7.5. A *vertex cover* for a graph $G = (V, E)$ is a subset V' of V such that every edge in E is incident with a vertex in V'. The vertex cover problem is: Given G and a positive integer ℓ, does G have a vertex cover with ℓ vertices? Show that the clique problem (see Exercise 7.4) is polynomial reducible to the vertex cover problem.

7.6. Give a necessary and sufficient condition for a graph to be colorable with one color.

7.7. Write an algorithm to determine whether a graph $G = (V, E)$ is 2-colorable. If your algorithm does more than $\Theta(\max\{n, m\})$ operations where $n = |V|$ and $m = |E|$, then try again.

7.8. Devise an algorithm to determine the chromatic number of graphs with the property that each vertex has degree at most two (i.e., is incident with at most two edges). The running time of your algorithm should be linear in the number of vertices in the graph.

7.9. Devise a polynomial-bounded algorithm to determine whether a CNF expression with at most two literals per clause is satisfiable. What is the worst-case complexity of your algorithm?

7.10. It is known that the largest clique possible in a planar graph has four vertices. Devise a polynomial-bounded algorithm to determine if a graph has a 4-clique. What is the worst-case complexity of your algorithm?

7.11. Give necessary and sufficient conditions for a graph (undirected) with maximum degree 2 to have a Hamilton circuit. Outline an efficient algorithm to test the conditions.

7.12. Consider the following problem:

You are given a packing of $s_1, \ldots, s_n$ into k bins, i.e., a partition of $\{1, 2, \ldots, n\}$ into k subsets $T_1, \ldots, T_k$ such that $\Sigma_{i \in T_j} s_i \leqslant 1$ for $1 \leqslant j \leqslant k$. Determine whether or not the s_i can be packed in $k - 1$ bins.

Show that this problem is $\mathcal{NP}$-complete.

Section 7.2: Approximation Algorithms

7.13. We may state the CNF-satisfiability problem as an optimization problem in the following form:

Given a CNF expression E, find a truth assignment for the variables in E to make the maximum possible number of clauses true.

Describe the set FS_E and the function v_E.

7.14. Consider the following problem:

Let $F = \{S_1, \ldots, S_n\}$ be a set of subsets of a set A such that $\cup_{i=1}^{n} S_i = A$. Find a minimal subset of F, say $\{S_{i_1}, \ldots, S_{i_k}\}$ that covers A, i.e., such that $\cup_{j=1}^{k} S_{i_j} = A$.

Describe the set FS_F and the function v_F.

Section 7.3: The Knapsack Problem

7.15 Devise an algorithm that, when given n and k such that $1 \leqslant k \leqslant n$, generates all subsets of $\{1, 2, \ldots, n\}$ containing at most k elements. The number of operations done between generating one subset and generating the next one should be $O(k)$ and independent of n.

7.16. Extend the approximation algorithms for the knapsack problem and Theorems 7.6 and 7.7 to the more general formulation of the problem given at the end of Section 7.3.

Section 7.4: Bin Packing

7.17. a) Construct an example for the bin packing problem where the NIFF algorithm uses three bins but the optimal number is two.
   b) Construct an infinite sequence of examples I_t where I_t has n_t objects and $n_1 < n_2 < \cdots$, and $v^*(I_t) = 2$ but NIFF uses three bins.

7.18. Show that Lemma 7.9 cannot be made stronger by constructing a sequence of examples such that for each $k \geqslant 2$ there is an input I with $v^*(I) = k$ and NIFF puts $k - 1$ objects in extra bins.

7.19. Show that if $2 \leqslant v^*(I) \leqslant 4$, NIFF uses at most $v^*(I) + 1$ bins.

7.20. Construct a sequence of inputs I_t for the bin packing problem such that $v^*(I_t) = v_t, v_1 < v_2 < \cdots$, and such that $r_{\text{NIFF}}(I_t) \geqslant \frac{7}{6}$ for all t.

7.21. Write a NONINCREASING BEST FIT algorithm for bin packing. What is the order of the worst-case running time?

7.22. a) Write a NEXT FIT algorithm for bin packing.
   b) Prove that $r_{\text{NF}}(I) \leqslant 2$ for all inputs I.

7.23. Consider the following problem:

You have t bins, each with unbounded capacity, and are given $s_1, \ldots, s_n$. Pack the s_i into the bins so as to minimize the maximum bin level.

This problem may be thought of as a job scheduling problem where the bins are processors and the s_i are the time requirements for n independent jobs. The problem is to assign jobs to processors to minimize the total time required. Write an approximation algorithm A for this problem. Indicate what its worst-case time requirement is and say as much as you can about the quality of its output, i.e., about the functions S_A and R_A.

Section 7.5: Graph Coloring

7.24. Describe the data structures you would use in Algorithm 7.3 to represent the graph and the coloring. Rewrite the algorithm with the implementation details and show that the number of operations done is $O(n^2)$.

7.25. Describe how the SCI strategy behaves on the graphs G_k defined in Section 7.5. In particular, how many times are pairs of colors interchanged?

NOTES AND REFERENCES

Most of the material in Section 7.1 is described more formally in Karp (1972) and Aho, Hopcroft, and Ullman (1974). The former shows that the two characterizations of $\mathcal{NP}$ given here are equivalent, contains a long list of $\mathcal{NP}$-complete problems, and outlines proofs of polynomial reducibility among them. The latter contains the details of some of those proofs and a proof of Theorem 7.2 (which was originally proved in a different form in Cook (1971)). Garey, Johnson, and Stockmeyer (1974) examine the effects of restricting the set of inputs on the $\mathcal{NP}$-completeness of various problems. There are numerous other articles showing that various problems are $\mathcal{NP}$-complete; one source for such papers is the proceedings of the last several annual ACM symposia on Theory of Computing.

For definitions of several formal models of computation, see any of the many books on the theory of computation, e.g., Minsky (1967) or Brainerd and Landweber (1974). Chapter 1 in Aho, Hopcroft, and Ullman (1974), Chapter 10 in Hopcroft and Ullman (1969), and Cook and Reckhow (1973) specifically consider how the complexity of problems differ on various models of computation and prove the invariance of $\mathcal{P}$ with respect to several models.

The approximation algorithms described in Sections 7.2–7.5 are from Sahni (1975) (knapsack), Garey, Graham, and Ullman (1972) and Johnson (1972) (bin packing), and Johnson (1974) (graph coloring). See Johnson (1973) and Sahni (1976) for approximation algorithms for other problems. There are numerous other papers on approximation algorithms. The results mentioned at the end of Section 7.5 and other theorems concerning the unlikeliness of obtaining good approximation algorithms for some problems are proved in Garey and Johnson (1976) and Sahni and Gonzalez (1976).

For general discussion and some additional references, see Weide (1977).

Bibliography

Aho, A.V., and M.J. Corasick (1975). "Efficient string matching: An aid to bibliographic search." *Communications of the ACM* 18: 333–340.

Aho, A.V., J.E. Hopcroft, and J.D. Ullman (1974). *The Design and Analysis of Computer Algorithms.* Reading, Mass.: Addison-Wesley.

Angluin, D. (1976). "The four Russians' algorithm for Boolean matrix multiplication is optimal in its class." *SIGACT News* 8 (No. 1): 29–33.

Arlazarov, V.L., E.A. Dinic, M.A. Kronrod, and I.A. Faradzev (1970). "On economical construction of the transitive closure of a directed graph." *Soviet Mathematics, Doklady* 11 (No. 5): 1209–1210.

Borodin, A., and I. Munro (1975). *Computational Complexity of Algebraic and Numeric Problems.* New York: American Elsevier.

Boyer, R.S., and J.S. Moore (1977). "A fast string searching algorithm." *Communications of the ACM* 20 (No. 10): 762–772

Brainerd, W.S., and L.H. Landweber (1974). *Theory of Computation.* New York: John Wiley.

Brigham, E.O. (1974). *The Fast Fourier Transform.* Englewood Cliffs, N.J.: Prentice-Hall.

Cook, S.A. (1971). "The complexity of theorem proving procedures." *Proceedings of the Third Annual ACM Symposium on Theory of Computing,* pp. 151–158.

Cook, S.A., and R.A. Reckhow (1973). "Time-bounded random access machines." *Journal of Computer and System Sciences* 7: 354–375.

Cooley, J.W., and J.W. Tukey (1965). "An algorithm for the machine calculation of complex Fourier series." *Mathematics of Computation* 19: 297–301.

Dahl, O.-J., E.W. Dijkstra, and C.A.R. Hoare (1972). *Structured Programming.* Washington, D.C.: Academic Press.

Day, A.C. (1972). *Fortran Techniques.* New York: Cambridge University Press.

Deo, N. (1974). *Graph Theory with Applications to Engineering and Computer Science.* Englewood Cliffs, N.J.: Prentice-Hall.

Dijkstra, E.W. (1959). "A note on two problems in connexion with graphs." *Numerische Mathematik* **1**: 269–271.

Elspas, B., K.N. Levitt, R.J. Waldinger, and A. Waksman (1972). "An assessment of techniques for proving program correctness." *ACM Computing Surveys* **4** (No. 2): 97–147.

Even, S. (1973) *Algorithmic Combinatorics*. New York: Macmillan.

Fischer, M.J. (1972). "Efficiency of equivalence algorithms." In R.E. Miller and J.W. Thatcher (eds.), *Complexity of Computer Computations*, pp. 153–167. New York: Plenum Press.

Fischer, M.J., and A.R. Meyer (1971). "Boolean matrix multiplication and transitive closure." *Conference Record, IEEE 12th Annual Symposium on Switching and Automata Theory*, pp. 129–131.

Ford, L.R., Jr., and D.R. Fulkerson (1962). *Flows in Networks*. Princeton, N.J.: Princeton University Press.

Gabow, H.N. (1977). "Two algorithms for generating weighted spanning trees in order." *SIAM Journal on Computing* **6** (No. 1): 139–150.

Galil, Z. (1976). "Real-time algorithms for string-matching and palindrome recognition." *Proceedings of the Eighth Annual ACM Symposium on Theory of Computing*, pp. 161–173.

Galler, B.A., and M.J. Fischer (1964). "An improved equivalence algorithm." *Communications of the ACM* **7** (No. 5): 301–303.

Garey, M.R., R.L. Graham, and J.D. Ullman (1972). "Worst-case analysis of memory allocation algorithms." *Proceedings of the Fourth Annual ACM Symposium on Theory of Computing*, pp. 143–150.

Garey, M.R., and D.S. Johnson (1976). "The complexity of near-optimal graph coloring." *Journal of the ACM* **23** (No. 1): 43–49.

Garey, M.R., D.S. Johnson, and L. Stockmeyer (1974). "Some simplified $\mathcal{NP}$-complete problems." *Proceedings of the Sixth Annual ACM Symposium on Theory of Computing*, pp. 47–63.

Gentleman, W.M., and G. Sande (1966). "Fast Fourier transforms – for fun and profit." *Proceedings, Fall Joint Computer Conference*, pp. 563–578.

Good, I.J. (1968). "A five-year plan for automatic chess." In E. Dale and D. Michie (eds.), *Machine Intelligence*, Volume 2, pp. 89–118. New York: American Elsevier.

Graham, R.L. (1969). "Bounds on multiprocessing timing anomalies." *SIAM Journal of Applied Math* **17** (No. 2): 416–429.

Hantler, S.L., and J.C. King (1976). "An introduction to proving the correctness of programs." *ACM Computing Surveys* **8** (No. 3): 331–353.

Harary, F. (1969). *Graph Theory*. Reading, Mass.: Addison-Wesley.

Hopcroft, J.E., and R.E. Tarjan (1973a). "Dividing a graph into triconnected components." *SIAM Journal on Computing* **2** (No. 3): 135–157.

Hopcroft, J.E., and R.E. Tarjan (1973b). "Efficient algorithms for graph manipulation." *Communications of the ACM* 16 (No. 6): 372–378.

Hopcroft, J.E., and R.E. Tarjan (1974). "Efficient planarity testing." *Journal of the ACM* 21 (No. 4): 549–568.

Hopcroft, J.E., and J.D. Ullman (1969). *Formal Languages and Their Relation to Automata.* Reading, Mass.: Addison-Wesley.

Hopcroft, J.E., and J.D. Ullman (1973). "Set merging algorithms." *SIAM Journal on Computing* 2 (No. 4): 294–303.

Horowitz, E., and S. Sahni (1974). "Computing partitions with applications to the knapsack problem." *Journal of the ACM* 21 (No. 2): 277–292.

Horowitz, E., and S. Sahni (1976). *Fundamentals of Data Structures.* Woodland Hills, Calif.: Computer Science Press.

Johnson, D.S. (1972). "Fast allocation algorithms." *Proceedings of the Thirteenth Annual Symposium on Switching and Automata Theory*, pp. 144–154.

Johnson, D.S. (1973). "Approximation algorithms for combinatorial problems." *Proceedings of the Fifth Annual ACM Symposium on Theory of Computing,* pp. 38–49.

Johnson, D.S. (1974). "Worst-case behavior of graph coloring algorithms." *Proceedings of the Fifth Southeastern Conference on Combinatorics, Graph Theory, and Computing*, pp. 513–528. Winnipeg, Canada: Utilitas Mathematica Publishing.

Karp, R.M. (1972). "Reducibility among combinatorial problems." In R.E. Miller and J.W. Thatcher (eds.), *Complexity of Computer Computations*, pp. 85–104. New York: Plenum Press.

Knuth, D.E. (1968). *The Art of Computer Programming, Volume I: Fundamental Algorithms.* Reading, Mass.: Addison-Wesley.

Knuth, D.E. (1973). *The Art of Computer Programming, Volume III: Sorting and Searching.* Reading, Mass.: Addison-Wesley.

Knuth, D.E. (1976). "Big omicron and big omega and big theta." *SIGACT News* 8 (No. 2): 18–24.

Knuth, D.E. (1977). "The complexity of songs." *SIGACT News* 9 (No. 2): 17–24.

Knuth, D.E., J.H. Morris, Jr., and V.R. Pratt (1977). "Fast pattern matching in strings." *SIAM Journal on Computing* 6 (No. 2): 240–267.

Minsky, M. (1967). *Computation: Finite and Infinite Machines.* Englewood Cliffs, N.J.: Prentice-Hall.

Morris, J.H., Jr., and V.R. Pratt (1970). "A linear pattern-matching algorithm." *Tech. Rep.* 40. University of California, Berkeley.

Nievergelt, J., and J.C. Farrar (1972). "What machines can and cannot do." *ACM Computing Surveys* 4 (No. 2): 81–96.

Rabin, M.O. (1977). "Complexity of computations." *Communications of the ACM* 20 (No. 9): 625–633.

Reingold, E.M. (1972). "On the optimality of some set merging algorithms." *Journal of the ACM* **19** (No. 4): 649–659.

Reingold, E.M., J. Nievergelt, and N. Deo (1977). *Combinatorial Algorithms: Theory and Practice*. Englewood Cliffs, N.J.: Prentice-Hall.

Reingold, E.M., and A.I. Stocks (1972). "Simple proofs of lower bounds for polynomial evaluation." In R.E. Miller and J.W. Thatcher (eds.), *Complexity of Computer Computations*, pp. 21–30. New York: Plenum Press.

Sahni, S. (1975). "Approximate algorithms for the 0/1 knapsack problem." *Journal of the ACM* **22** (No. 1): 115–124.

Sahni, S.K. (1976). "Algorithms for scheduling independent tasks." *Journal of the ACM* **23** (No. 1): 116–127.

Sahni, S.K., and T. Gonzalez (1976). "$\mathscr{P}$-complete approximation problems." *Journal of the ACM* **23** (No. 3): 555–565.

Savage, J.E. (1974). "An algorithm for the computation of linear forms." *SIAM Journal on Computing* **3**: 150–158.

Savage, J.E. (1976). *The Complexity of Computing*. New York: John Wiley.

Sedgewick, R. (1977). "Quicksort with equal keys." *SIAM Journal on Computing* **6** (No. 2): 240–267.

Strassen, V. (1969). "Gaussian elimination is not optimal." *Numerische Mathematik* **13**: 354–356.

Tarjan, R.E. (1972). "Depth-first search and linear graph algorithms." *SIAM Journal on Computing* **1** (No. 2): 146–160.

Tarjan, R.E. (1975). "On the efficiency of a good but not linear set union algorithm." *Journal of the ACM* **22** (No. 2): 215–225.

Warshall, S. (1962). "A theorem on Boolean matrices." *Journal of the ACM* **9** (No. 1): 11–12.

Wegner, P. (1974). "Modification of Aho and Ullman's correctness proof of Warshall's algorithm." *SIGACT News* **6** (No. 1): 32–35.

Weide, B. (1977). "A survey of analysis techniques for discrete algorithms." *ACM Computing Surveys* **9** (No. 4): 291–313.

Winograd, S. (1970). "On the number of multiplications necessary to compute certain functions." *Journal of Pure and Applied Math* **23**: 165–179.

Yao, A.C. (1976). "On the average behavior of set merging algorithms." *Proceedings of the Eighth Annual ACM Symposium on Theory of Computing*. pp. 192–195.

Index

Folios in italic indicate figures.